Ignorance

Ignorance

How It Drives Science

Stuart Firestein

OXFORD
UNIVERSITY PRESS

OXFORD
UNIVERSITY PRESS

Oxford University Press, Inc., publishes works that further
Oxford University's objective of excellence
in research, scholarship, and education.

Oxford   New York
Auckland   Cape Town   Dar es Salaam   Hong Kong   Karachi
Kuala   Lumpur   Madrid   Melbourne   Mexico City   Nairobi
New Delhi   Shanghai   Taipei   Toronto

With offices in
Argentina   Austria   Brazil   Chile   Czech Republic   France   Greece
Guatemala   Hungary   Italy   Japan   Poland   Portugal   Singapore
South   Korea   Switzerland   Thailand   Turkey   Ukraine   Vietnam

Published by Oxford University Press, Inc.
198 Madison Avenue, New York, New York 10016
www.oup.com

Library of Congress Cataloging-in-Publication Data
Firestein, Stuart.
Ignorance : how it drives science/Stuart Firestein.
p. cm.
Includes bibliographical references and index.
ISBN 978–0–19–982807–4 (hardback)
1. Science–Philosophy.   2. Ignorance (Theory of knowledge)
3. Discoveries in science.   I. Title.
Q175.32.K45F57 2012
501'.9—dc23
2011051395

5 7 9 8 6 4
Printed in United States of America
on acid-free paper

Acknowledgments

A t the start of my *Ignorance* class, in response to the inevitable questions about grades, I warn the students, partially tongue in cheek, that they should consider carefully what grade they would actually like to appear on their records. After all, the transcript will read "SNC3429 *Ignorance*"—and do they want the grade that appears after this to be an A...or an F? There is some of the same uneasiness about acknowledging the many contributions of friends, colleagues, students, and family to a book called *Ignorance*. Nonetheless, my debt is great, and I can only hope that these many co-conspirators will be happy to have their names mentioned here. Special thanks then to the many wonderful students at Columbia

University who gambled on taking a class called *Igno-rance* and have added so much to it. Teaching this class has been one of the real highlights of my career at the University. And of course there are my courageous col-leagues, working scientists who have taken two hours out of an evening and bravely put their ignorance on display, captivating and enlightening both the students and me. Some of them appear in the case histories that are part of this book, and the names of the others can be found at the Ignorance Web site. I had the tremendous fortune to have a member of my laboratory volunteer to be a teaching assistant at the outset of the class and he helped me develop it intellectually, and in many other ways that insured its success. His name is Alex Chesler and you will hear from him, I'm sure. After Alex left, Isabel Gabel served as a teaching assistant and what was so special about Isabel was that she was a graduate stu-dent in the History Department and so brought quite a different and fresh perspective.

Several close colleagues and friends, scientists and humanists, have read various versions of this manu-script and have been extremely generous, not to men-tion unflinching, in their comments. They include Terry Acree, Charles Greer, Christian Margot, Patrick Fitzger-ald, Peter Mombaerts, Philip Kitcher, Cathy Popkin, Gordon Shepherd, Jonathan Weiner, and Nick Hern.

Many important things have been changed according to their critiques, but any foolishness that remains is entirely my responsibility.

Three years ago a small group of neuroscience graduate students and graduate MFA nonfiction writing students here at Columbia came to me with a proposal to form a writing group with the particular purpose of putting our heads together and trying to learn how to write about real science for a real public audience. This group has been dubbed Neuwrite, although our subjects often wander far afield of neuroscience. Sections of this book have been mercilessly workshopped by this remarkable and talented group, and it is impossible for me to overstate what I have learned through the generosity and insight of this band of young writers.

I have been fortunate to have several editors who have been not just supportive but truly enthusiastic about this project. First Catherine Carlin, who suggested making a book from the class, and more recently Joan Bossert, who has taken over this manuscript as though it were a child. It was Joan who asked Marion Osmun to perform a serious edit on an early draft and who so thoroughly got what the book was about. The Alfred P Sloan Foundation supports a program to make science accessible to the public, and they have been generous in supporting this project with a grant. I should also note that an early proponent of the importance of ignorance in science was a former director of the Foundation, Frank Gomoroy.

And mostly my greatest debt of gratitude is to my wife, Diana, and daughter, Morgan, who have shown unshakeable faith in my ignorance, and other important things, for as long as they have known me.

Contents

Ignorance

Introduction

"It is very difficult to find a black cat in a dark room," warns an old proverb. "Especially when there is no cat."

This strikes me as a particularly apt description of how science proceeds on a day-to-day basis. It is certainly more accurate than the more common metaphor of scientists patiently piecing together a giant puzzle. With a puzzle you see the manufacturer has guaranteed there is a solution.

I know that this view of the scientific process—feeling around in dark rooms, bumping into unidentifiable things, looking for barely perceptible phantoms—is contrary to that held by many people, especially by nonscientists. When most people think of science, I suspect they imagine the nearly 500-year-long systematic pursuit of knowledge that, over 14 or so generations, has uncovered more information about the universe and everything in it than all that was known in the first 5,000 years of recorded human history. They imagine a brotherhood tied together by its golden rule, *the Scientific Method*, an immutable set of precepts for devising experiments that churn out the cold, hard facts. And these solid facts form the edifice of science, an unbroken record of advances and insights embodied in our modern views and unprecedented standard of living. Science, with a capital *S*.

That's all very nice, but I'm afraid it's mostly a tale woven by newspaper reports, television documentaries, and high school lesson plans. Let me tell you my somewhat different perspective. It's not facts and rules. It's black cats in dark rooms. As the Princeton mathematician Andrew Wiles describes it: It's groping and probing and poking, and some bumbling and bungling, and then a switch is discovered, often by accident, and the light is lit, and everyone says, "Oh, wow, so that's how it looks," and then it's off into the next dark room, looking for the next mysterious black feline. If this all sounds depressing, perhaps some bleak Beckett-like scenario of existential endlessness, it's not. In fact, it's somehow exhilarating.

This contradiction between how science is pursued versus how it is perceived first became apparent to me in my dual role as head of a laboratory and Professor of Neuroscience at Columbia University. In the lab, pursuing questions in neuroscience with the graduate students and postdoctoral fellows, thinking up and doing experiments to test our ideas about how brains work, was exciting and challenging and, well, exhilarating. At the same time I spent a lot of time writing and organizing lectures about the brain for an undergraduate course that I was teaching. This was quite difficult given the amount of information available, and it also was an interesting challenge. But I have to admit it was not exhilarating. What was the difference?

The course I was, and am, teaching has the forbidding-sounding title "Cellular and Molecular Neuroscience." The students who take this course are very bright young people in their third or fourth year of University and are mostly declared biology majors. That is, these students are all going on to careers in medicine or biological research. The course consists of 25 hour-and-a-half lectures and uses a textbook with the lofty title *Principles of Neural Science*, edited by the eminent neuroscientists Eric Kandel and Tom Jessell (with the late Jimmy Schwartz). The textbook is 1,414 pages long and weighs in at a hefty 7.7 pounds, a little more in fact than twice the weight of a human brain. Now, textbook writers are in the business of providing more information for the buck than their competitors, so the books contain quite a lot of detail. Similarly, as a lecturer, you wish to sound authoritative, and you want your lectures to be "informative," so you tend to fill them with many facts hung loosely on a few big concepts. The result, however, was that by the end of the semester I began to sense that the students must have had the impression that pretty much everything is known in neuroscience. This couldn't be more wrong. I had, by teaching this course diligently, given these students the idea that science is an accumulation of facts. Also not true. When I sit down with colleagues over a beer at a meeting, we don't go over the facts, we don't talk about what's known; we talk about what we'd like to figure out, about what needs to be done. In a letter to her brother in 1894, upon having just received her *second* graduate degree, Marie Curie wrote: "One never notices what has been done; one can only see what remains to be done ..."

This crucial element in science was being left out for the students. The undone part of science that gets us into the lab early and keeps us there late, the thing that "turns your crank," the very driving force of science, the exhilaration of the unknown, all this is missing from our classrooms. In short, we are failing to teach the *ignorance*, the most critical part of the whole operation.

And so it occurred to me that perhaps I should mention some of what we don't know, what we still need to find out, what are still mysteries, what still needs to be done—so that these students can get out there and find out, solve the mysteries and do these undone things. That is, I should teach them ignorance. Finally, I thought, a subject I can excel in.

This curious revelation grew into an idea for an entire course devoted to, and titled, Ignorance. A science course. That course, in its current incarnation, began in the spring of 2006. At the heart of the course are sessions, I hesitate to call them classes, in which a guest scientist talks to a group of students for a couple of hours about what he or she doesn't know. They come and tell us about what they would like to know, what they think is critical to know, how they might get to know it, what will happen if they do find this or that thing out, what might happen if they don't. About what could be known, what might be impossible to know, what they didn't know 10 or 20 years ago and know now, or still don't know. Why they want to know this and not that, this more than that. In sum, they talk about the current state of their ignorance.

Recruiting my fellow scientists to do this is always a little tricky—"Hello, Albert, I'm running a course on ignorance and I think you'd be perfect." But in fact almost every scientist realizes immediately that he or she would indeed be perfect, that this is truly what they do best, and once they get over not having any slides prepared for a talk on ignorance, it turns into a surprising and satisfying adventure. Our faculty has included astronomers, chemists, ecologists, ethologists, geneticists, mathematicians, neurobiologists, physicists, psychobiologists, statisticians, and zoologists. The guiding principle behind this course is not simply to talk about the *big questions*—how did the universe begin, what is consciousness, and so forth. These are the things of popular science programs like *Nature* or *Discovery*, and, while entertaining, they are not really *about* science, not the day-to-day, nitty-gritty, at the office and bench kind of science. Rather, this course aims to be a series of case studies of ignorance—the ignorance that drives science. In fact, I have taken examples from the class and presented them as a series of "case histories" that make up the second half of this book. Despite them being about people doing highly esoteric scientific work, I think you will find them engaging and pleasantly accessible narratives.

Now I use the word *ignorance* at least in part to be intentionally provocative. But let's take a moment to define the kind of ignorance I am referring to, because ignorance has many bad connotations, especially in common usage, and I don't mean any of those. One kind of ignorance is willful stupidity; worse than simple stupidity, it is a callow indifference to facts or logic. It shows itself as a stubborn devotion to uninformed opinions, ignoring (same root) contrary ideas, opinions, or data. The ignorant are unaware, unenlightened, uninformed, and surprisingly often occupy elected offices. We can all agree that none of this is good.

But there is another, less pejorative sense of ignorance that describes a particular condition of knowledge: the absence of fact, understanding, insight, or clarity about something.

It is not an individual lack of information but a communal gap in knowledge. It is a case where data don't exist, or more commonly, where the existing data don't make sense, don't add up to a coherent explanation, cannot be used to make a prediction or statement about some thing or event. This is knowledgeable ignorance, perceptive ignorance, insightful ignorance. It leads us to frame better questions, the first step to getting better answers. It is the most important resource we scientists have, and using it correctly is the most important thing a scientist does. James Clerk Maxwell, perhaps the greatest physicist between Newton and Einstein, advises that "Thoroughly conscious ignorance is the prelude to every real advance in science."

. . .

Before diving into all that ignorance, let me take a moment to provide a guide for reading this book. First, it's short, which you will have already noticed from its size. I would have liked it to be shorter, but as Pascal once said by way of apology at the end of a long note written to a friend, "I would have been briefer if I'd had more time." I would have been briefer if I'd been smarter, but this will have to do.

I have envisioned a reader who is not an expert. That, of course, includes everyone, since, in a field not our own, we are all beginners. Working scientists will, I believe, find much here that is familiar but rarely talked about; non-scientists will find a way to understand what may seem most baffling about science. It is this second reader that I especially care about and whom the book is largely written for and to.

I like to think of this book being read in one or two sittings, a couple of hours spent profitably focusing your mind on a perhaps novel way of thinking about science, and by extension other kinds of knowledge as well. The point is that it ought not to interfere with your daily life, your occupation, your work, by exacting a significant debit on your valuable time. It should add to it.

To accomplish this, I have taken several steps aimed at making the book easier to scan. I have not included extensive and distracting notes, citations, or footnotes. Where someone is quoted in the text, and their identity is obvious, I have not added any further citation material—you can look these things up on the Web easily enough. Where further notes or expansion of the material could provide something interesting to some readers but are not integral to the forward progress of the text, I have included suggested reading at the end with comments and often keyed to particular points in the text. There is a Web site—http://ignorance. biology.columbia.edu—for the book, and the course that it is based on, with a great deal more information on it, and this will be available to interested readers.

The form of the book is also aimed at producing a reasonably manageable read. The book is split into two distinct sections. The first half is an essay form and the second half is narrative, composed of four case histories of ignorance, that I think you will find engaging and revealing, based on the classes in my course. In the essay portion what I hope are a few crucial ideas are repeated in slightly different ways from different angles to add up to a novel perspective. I have learned from years of teaching that saying nearly the same thing in different ways is an often effective strategy. Sometimes a person has to hear something a few times or just the right way to get that click of recognition, that "ah-ha moment" of clarity. And even if you completely get it the first time, another explanation always adds texture. Thus, it is not a "well-organized" book in the sense that the chapters lead the reader through a thicket of facts and concepts to an inescapable conclusion. It is not so much a discourse, as a musing with a point. I have considered various ways of organizing the material and what appears here is to my mind the most straightforward, if not the most attractive. I invite the reader to wander through the material rather than be guided down a path of argument.

ONE

A Short View of Ignorance

Knowledge is a big subject. Ignorance is bigger. And it is more interesting.

Perhaps this sounds strange because we all seek knowledge and hope to avoid ignorance. We want to know how to do this, and get that, and succeed in various endeavors. We go to school for many years, in some cases now for more than 20 years of formal schooling, often followed by another 4–8 years of "on-the-job" training in internships, fellowships, residencies, and the like—all to gain more knowledge. But how many of us think about what comes after the knowledge is acquired? We may spend 20-plus years being educated, but what about the following 40 years? For those years we foolishly have no well-defined program, and much of the time we do not even have an inkling of what to do with them. So what does come *after* knowledge? You might not think of it in this order, but I would say that ignorance follows knowledge, not the other way around.

On her way into life-threatening surgery, Gertrude Stein was asked by her lifelong companion, Alice B. Toklas, "What is the answer?" Stein replied, "What is the question?" There are a few different versions of this story, but they all come to the same thing: Questions are more relevant than answers. Questions are bigger than answers. One good question can give rise to several layers of answers, can inspire decades-long searches for solutions, can generate whole new fields of inquiry, and can prompt changes in entrenched thinking. Answers, on the other hand, often end the process.

Are we too enthralled with the answers these days? Are we afraid of questions, especially those that linger too long? We seem to have come to a phase in civilization marked by a voracious appetite for knowledge, in which the growth of information is exponential and, perhaps more important, its availability easier and faster than ever. Google is the symbol, the insignia, the coat of arms of the modern world of information. More information is demanded, more facts are offered, more data are requested, and more is delivered more quickly. According to the Berkeley Institute, in the year 2002, 5 exabytes of information were added to the world's stores. That's a billion billion bits of data, enough to fill the Library of Congress 37,000 times over. This means 80 megabytes for every individual on the planet, equaling a stack of books 30 feet high for each of us to read. That was in 2002. It appears to have increased by a million times according to the latest update in this series for 2007.

What can one do in the face of this kind of information growth? How can anyone hope to keep up? How come we have not ground to a halt in the deepening swamp of information? Would you be suspicious if I told you it was just a matter of perspective? Working scientists don't get bogged down in the factual swamp because they don't care all that much for facts. It's not that they discount or ignore them, but rather that they don't see them as an end in themselves. They don't stop at the facts; they begin there, right beyond the facts, where the facts run out. Facts are selected, by a process that is a kind of controlled neglect, for the questions they create, for the ignorance they point to. What if we cultivated ignorance instead of fearing it, what if we controlled neglect instead of feeling guilty about it, what if we understood the power of *not* knowing in a world dominated by information? As the first philosopher, Socrates, said, "I know one thing, that I know nothing."

Scholars agree that Isaac Newton, in 1687, formulating the laws of force and inventing the calculus in his *Principia*

Mathematica, probably knew all of the extant science at that time. A single human brain could know everything there was to know in science. Today this is clearly impossible. Although the modern high school student probably possesses more scientific information than Newton did at the end of the 17th century, the modern professional scientist knows a far, far smaller amount of the available knowledge or information at the beginning of the 21st century. Curiously, as our collective knowledge grows, our ignorance does not seem to shrink. Rather, we know an ever smaller amount of the total, and our individual ignorance, as a ratio of the knowledge base, grows. This ignorance is a kind of limit, and it's frankly a bit annoying, at least to me, because the one thing you know is that there is so much more out there that you will never know. Unfortunately, there seems to be nothing that can be done about this.

On the grander scale there is absolute or true ignorance, the ignorance represented by what really isn't known, by anybody, anywhere—that is, communal ignorance. And this ignorance, the still mysterious, is also increasing. In this case, however, that's the good news, because it's not a limit; it is an opportunity. A Google search on the word "ignorance" gives 37 million hits; one on "knowledge" returns 495 million. This reflects Google's utility but also its prejudice. Surely there is more ignorance than knowledge. And because of that there is more left to do.

I feel better about all that ignorance than I do about all that knowledge. The vast archives of knowledge seem impregnable, a mountain of facts that I could never hope to learn, let alone remember. Libraries are both awe inspiring and depressing. The cultural effort that they represent, to record over generations what we know and think about the world and ourselves, is unquestionably majestic; but the impossibility of reading even a small fraction of the books inside them can be personally dispiriting.

Nowhere is this dynamic more true than in science. Every 10–12 years there is an approximate doubling of the number of scientific articles. Now this is not entirely new—it's actually been going on since Newton—and scientists have been complaining about it for almost as long. Francis Bacon, the pre-Enlightenment father of the scientific method, complained in the 1600s of how the mass of accumulated knowledge had become unmanageable and unruly. It was perhaps the impetus for the Enlightenment fascination with classification and with encyclopedias, an attempt to at least alphabetize knowledge, if not actually contain it. And the process is exponential, so it gets "worser and worser," as they say, over time. That first doubling of information amounted to a few tens of new books or papers, while the most recent doubling saw more than 1,000,000 new publications. It's not just the rate of increase; it's the actual amount that makes the pile so daunting. How does anyone even get started being a scientist? And if it's intimidating to trained and experienced scientists, what could it be to the average citizen? No wonder science attracts only the most devoted. Is this the reason that science seems so inaccessible?

Well, it is difficult, and there is no denying that there are a lot of facts that you have to know to be a professional scientist. But clearly you can't know all of them, and knowing lots of them does not automatically make you a scientist, just a geek. There are a lot of facts to be known in order to be a professional anything—lawyer, doctor, engineer, accountant, teacher. But with science there is one important difference. The facts serve mainly to access the ignorance. As a scientist, you don't do something with what you know to defend someone, treat someone, or make someone a pile of money. You use those facts to frame a new question—to speculate about a new black cat. In other words, scientists don't concentrate on what they know, which is considerable but also miniscule, but rather on what they don't know. The one big fact is that science traffics in ignorance, cultivates it, and is driven by it. Mucking about in the unknown is an adventure; doing it for a living is something most scientists consider a privilege. One of the crucial ideas of this book is that ignorance of this sort need not be the province of scientists alone, although it must be admitted that the good ones are the world's experts in it. But they don't own it, and you can be ignorant, too. Want to be on the cutting edge? Well, it's all, or mostly, ignorance out there. Forget the answers, work on the questions.

In the early days of television, the pioneering performer Steve Allen introduced on his variety show a regular routine known as The Question Man. The world it seemed had an overabundance of answers but too few questions. In the postwar 1950s, with its emphasis on science and technology, it could easily have felt this way to many people. The Question Man would be given an answer, and it was his task to come up with the question. We need The Question Man again. We still have too many answers, or at least we put too much stock in answers. Too much emphasis on the answers and too little attention to the questions have produced a warped view of science. And this is a pity, because it is the questions that make science such a fun game.

But surely all those facts must be good for something. We pay a very high price for them, in both money and time, and one hopes they are worth it. Of course, science creates and uses facts; it would be foolish to pretend otherwise. And certainly to be a scientist you have to know these facts or some subset of them. But how does a scientist *use* facts beyond simply accumulating them? As raw material, not as finished product. In those facts is the next round of questions, improved questions with new unknowns. Mistaking the raw material for the product is a subtle error but one that can have surprisingly far-reaching consequences.

Understanding this error and its ramifications, and setting it straight, is crucial to understanding science.

The poet John Keats hit upon an ideal state of mind for the literary psyche that he called Negative Capability—"that is when a man is capable of being in uncertainties, Mysteries, doubts without any irritable reaching after fact & reason." He considered Shakespeare to be the exemplar of this state of mind, allowing him to inhabit the thoughts and feelings of his characters because his imagination was not hindered by certainty, fact, and mundane reality (think Hamlet). This notion can be adapted to the scientist who really should always find himself or herself in this state of "uncertainty without irritability." Scientists do reach after fact and reason, but it is when they are most uncertain that the reaching is often most imaginative. Erwin Schrodinger, one of the great philosopher-scientists, says, "In an honest search for knowledge you quite often have to abide by ignorance for an indefinite period." (Schrodinger knew something about uncertainty; he posed the now famous Schrodinger's cat thought experiment in which a cat placed in a box with a vial of poison that could or could not be activated according to some quantum event was, until observed, both dead and alive, or neither.) Being a scientist requires having faith in uncertainty, finding pleasure in mystery, and learning to cultivate doubt. There is no surer way to screw up an experiment than to be certain of its outcome.

To summarize, my purpose in this intentionally short book is to describe how science progresses by the growth of ignorance, to disabuse you of the popular idea that science is entirely an accumulation of facts, to show how you can be part of the greatest adventure in the history of human civilization without slogging through dense texts and long lectures. You won't be a scientist at the end of it (unless you're already one), but you won't have to feel as if you're excluded from participating in the remarkable worldview that science offers, if you want to. I'm not proselytizing for science as the only legitimate way to understand the world; it's clearly not that. Many cultures have lived, and continue to live, quite happily without it. But in a scientifically sophisticated culture, such as ours, it is as potentially dangerous for the citizenry to be oblivious about science as it is for them to be ignorant of finance or law. And aside from being a good citizen, it's simply too interesting and too much fun to ignore.

We might start by looking at how science gets its facts and at how that process is really one of ignorance generation. From there we can examine how scientists do their work—choosing and making decisions about their careers and the questions they will devote themselves to; how we teach or fail to teach science; and finally how nonspecialists can have access to science through the unlikely portal of ignorance.

TWO

Finding Out

Science, it is generally believed, proceeds by accumulating data through observations and manipulations and other similar activities that fall under the category we commonly call experimental research. The scientific method is one of observation, hypothesis, manipulation, further observation, and new hypothesis, performed in an endless loop of discovery. This is correct, but not entirely true, because it gives the sense that this is an orderly process, which it almost never is. "Let's get the data, and then we can figure out the hypothesis," I have said to many a student worrying too much about how to plan an experiment.

The purpose of experiments is of course to learn something. The words we use to describe this process are interesting. We say that some feature is revealed, we find something out, we discover something. In fact the word *discover* itself has an evocative literal meaning—"to *dis-cover*," that is, to uncover, to remove a veil that was hiding something already there, to reveal a fact. Some artists talk also of revealing or discovering as the basis of the creative act— Rodin claimed that his sculpting process was to remove the stone that was not part of the sculpture; Louis Armstrong said that the important notes were the ones he didn't play.

The direct result of this discovery process in science is data. Observations, measurements, findings, and results accumulate and at some point may gel into a fact. The literary critic and historian Mary Poovey recently wrote a noteworthy book titled *A History of the Modern Fact* in which she traces the development of the fact as a respected and preferred unit of knowledge. In its growth to this exalted position it has supposedly shed any debt to authority, opinion, bias, or perspective. That is, it can be trusted because it supposedly arose from unbiased observations and measurements without being affected by subjective interpretation. Obviously this is ridiculous, as she so exhaustively shows. No matter how objective the measurement, someone still had to decide to make that measurement, providing ample opportunity for bias to enter the scheme right there. And of course data and facts are always interpreted because they often fail to produce an uncontestable result. Nonetheless, this idealized view of the fact still commands a central place, especially in science education (although not so clearly in science practice), where facts occupy a position at least as exalted as truth, and where they provide credibility by being separated from opinion. Scientific facts are "disinterested," which certainly doesn't sound like much fun and may be why they have become so uninteresting.

I don't mean by all of this to demean facts, but rather to place them in a more accurate perspective, or at least in the perspective of the working scientist. Facts are what we work for in science, but they are not actually the currency of the community of scientists. It may seem surprising to the nonscientist, but all scientists know that it is facts that are unreliable. No datum is safe from the next generation of scientists with the next generation of tools. The known is never safe; it is never quite sufficient. And perhaps nonintuitively, the more exact the fact, the less reliable it is likely to be; a precise measurement can always be revised and made a decimal point more precise, a definitive prediction is more likely to be wrong than a vague one that allows several possible outcomes.

One of the more gratifying, if slightly indulgent, pleasures of actually doing science is proving someone wrong—even yourself at an earlier time. How do scientists even know for sure when they know something? When is something known to their satisfaction? When is the fact final?

In reality, only false science reveres "facts," thinks of them as permanent and claims to be able to know everything and predict with unerring accuracy—one might think here of astrology, for example. Indeed, when new evidence forces scientists to modify their theories, it is considered a triumph, not a defeat. Max Planck, the brilliant physicist who led the revolution in physics now known as quantum mechanics, was asked how often science changed. He replied: "with every funeral," a nod to the way science often changes on a generational time scale. As each new generation of scientists comes to maturity, unencumbered by the ideas and "facts" of the previous generation, conception and comprehension is free to change in ways both revolutionary and incremental. Real science is a revision in progress, always. It proceeds in fits and starts of ignorance.

THE DARK SIDE OF KNOWLEDGE

There are cases where knowledge, or apparent knowledge, stands in the way of ignorance. The luminiferous ether of late 19th-century physics is an example. This was the medium that was believed to permeate the universe, providing the substrate through which light waves could propagate. Albert Michelson was awarded a Nobel Prize in 1907 for failing to observe this ether in his experiments to measure the speed of light—possibly the only Nobel Prize awarded for an experiment that didn't work. He was also, as it happens, the first American to win a Nobel Prize. The ether was a black cat that had physicists measuring and testing and theorizing in a dark room for decades—until Michelson's experiments raised the specter that this particular feline didn't even exist, thereby allowing Albert Einstein to postulate a view of the universe in a new and previously unimaginable way with his theories of relativity.

Phrenology, the investigation of brain function through an analysis of cranial bumps, functioned as a legitimate science for nearly 50 years. Although it contained a germ of truth, certain mental faculties are indeed localized to regions of the brain, and many attested to its accuracy in describing personality traits, it is now clear that a large bump on the right side of your head just behind your ear has nothing to do with your being an especially combative person. Nonetheless, hundreds of scientific papers appeared in the literature, and several highly respected scientific names of the 19th century were attached to it. Charles Darwin, not himself a subscriber, was reckoned by an examination of a picture of his head to have "enough spirituality for ten priests"! In these, and many other cases (the magical *phlogiston* to explain combustion and rust, or the heat fluid *caloric*), apparent knowledge hid our ignorance and retarded progress. We may look at these quaint ideas smugly now, but is there any reason, really, to think that our modern science may not suffer from similar blunders? In fact, the more successful the fact, the more worrisome it may be. Really successful facts have a tendency to become impregnable to revision.

Here are two current examples:

- Almost everyone believes that the tongue has regional sensitivities—sweet is sensed on the tip, bitter on the back, salt and sour on the sides. Pictures of "tongue maps" continue to appear not only in popular books on taste and cooking but in medical textbooks as well. The only problem is that it's not true. The whole thing arose from the mistranslation of a German physiology textbook by a Professor D. P. Hanig, who claimed that his very anecdotal experiments showed that parts of the tongue were slightly more or slightly less sensitive to the four basic tastes. Very slightly as it has turned out when the experiments are done more carefully (you can try this on your own tongue with some salt and sugar, for example). The Hanig work was published in 1901 and the translation, which considerably overstated the findings and canonized the myth, was by the famed Harvard psychologist Edward G. Boring (joke to thousands of undergraduate psychology majors forced to read his textbooks) in 1942. Boring, by the way, was a pioneer in sensory perception who gave us the well-known ambiguous figure that can be seen, depending on how you look at it, as a beautiful young woman or an old hag. Perhaps at least in part because of Boring's stature the mythical tongue map was canonized into a fact and maintained by repetition, rather than experimentation, now having endured for more than a century.

- In the early 20th century, using what were then the new techniques of electrical recording from living tissue, two pioneering neuroscientists, Lord Adrian and Keffer Hartline, recorded electrical activity in the brain. The most prominent form of this activity was a train of brief pulses of voltage, typically less than a tenth of a volt and lasting only a few milliseconds. Adrian and Hartline characterized them as "spikes" because they appeared on their recording equipment as sharp vertical lines that looked like "spikes" of voltage. These spikes of voltage could appear singly or in trains that contained hundreds of spikes and could last for several seconds. Adrian recorded them in the cells that bring messages from the skin to the brain, and Hartline found them in cells in the retina. In both cases they noted that increases in the strength of the stimulus—touch or light—caused more rapid trains of spikes in these cells. These spikes have since been recorded in virtually every area of the brain and in all the sensory organs, and they have come to be regarded as the language of the brain—that is, they encode all the information passing into and around the brain. Spikes are a fundamental unit of neurobiology. For the last 75 years my neuroscience colleagues and I have been studying spikes and teaching our students about spikes and making grand theories about how the brain works based on spiking behavior. Some of it is true. But what have we missed by concentrating on spikes for the last eight decades? A lot, it turns out. There are many other electrical sorts of signals in the brain, not as prominent as spikes, but that's a reflection of our recording technology not of the brain itself. These other processes, as well as chemical events that are not electrical, and therefore can't even be seen with an electrical apparatus, are now being recognized as perhaps the more salient features of brain activity. But we have been mesmerized by spikes, and the rest has been virtually invisible, even though it is right in front of our faces, happening all the time in our brains. Spike analysis was a successful industry in neuroscience that occupied us for the better part of a century and filled journals and textbooks with mountains of data and facts. But it may have been too much of a good thing. We should also have been looking at what they didn't tell us about the brain.

I can't resist one more very recent example.

- For at least as long as I have been teaching neuroscience, I have told students that the human brain is composed of about 100 billion neurons and 10 times that number of glial cells—a kind of cell that nourishes neurons and provides some packing and structure for the organ (the word *glia* comes from Greek for "glue"). These numbers are also in all the textbooks. In early 2009 I received an e-mail from a neuroanatomist in Argentina named Suzana Herculano-Houzel asking me if I would help her group's research project by taking a short survey. Among the questions on that survey was how many neurons and glial cells I thought were in the human brain, and where I got that number from. The first part of the question was easy—I filled in the stock answers. But actually I wasn't sure where that number had come from. It was in the textbooks, but no references to any work were ever given for it. No one, it turned out, knew where the number came from. It sounded reasonable; it wasn't, after all, an exact number, not like 101,954,467,298 neurons, which would have required a reference to back it up. A little over a year later I heard back from Suzana. Her group had developed a new method for counting cells that was more exact and less prone to errors and could be used on big tissues, like brains. They counted the number of neurons and the number of glial cells in several human brains. For neurons they found that the average number for humans is 80 billion—20% less than we thought; and more remarkably for glial cells there were about an equal number as there were neurons—not 10 times more! In one fell swoop, we lost 120 billion cells in our brains! How could this have happened? How did that first, wrong number become so widespread? It seemed as though the text-book writers had just picked it up from one another and kept passing it around. The number became true as the result of repetition, not experiment.

WHAT SCIENCE MAKES

George Bernard Shaw, in a toast at a dinner feting Albert Einstein, proclaimed, "Science is always wrong. It never solves a problem without creating 10 more." Isn't that glorious? Science (and I think this applies to all kinds of research and scholarship) produces ignorance, possibly at a faster rate than it produces knowledge.

Science, then, is not like the onion in the often used analogy of stripping away layer after layer to get at some core, central, fundamental truth. Rather it's like the magic well: no matter how many buckets of water you remove, there's always another one to be had. Or even better, it's like the widening ripples on the surface of a pond, the ever larger circumference in touch with more and more of what's outside the circle, the unknown. This growing forefront is where science occurs. Curious then that in so many settings—the classroom, the television special, the newspaper accounts— it's the inside of the circle that seems so enticing, rather than what's out there on the ripple. It is a mistake to bob around in the circle of facts instead of riding the wave to the great expanse lying outside the circle. But that's still where most people who are not scientists find themselves.

Now it may seem obvious to say that science is about the unknown, but I would like to have a deeper look at this apparently simple statement, to see if we can't mine it for something more profound. Stories abound in the history of science of well-respected scientists claiming that everything but the measurements out to another decimal place were now known and all major outstanding questions were settled. At one time or another, geography, physics, chemistry, and so on were all declared finished. Obviously these claims were premature. Seems we don't always know what we don't know. In the inimitable words of no less than Donald H. Rumsfeld, the former US secretary of defense now best known for his incompetent handling of the war in Iraq, "there are known unknowns and unknown unknowns." He was roundly ridiculed for this and other tortured locutions, and in matters of war and security it might not be the clearest sort of thinking, but he was certainly right that there are things we don't even know we don't know. We might even go a step further and recognize that there are unknowable unknowns—things that we cannot know due to some inherent and implacable limitation. History, as a subject, could be said to be fundamentally unknowable; the data are lost and they are not recoverable.

So it's not so much that there are limits to our knowledge, more critically there may be limits to our ignorance. Can we investigate these limits? Can ignorance itself become a subject for investigation? Can we construct an epistemology of ignorance like we have one for knowledge? Robert Proctor, a historian of science at Stanford University, and perhaps best known as an implacable foe of the tobacco industry's misinformation campaigns, has coined the word *agnotology* as the study of ignorance. We can investigate ignorance with the same rigor as philosophers and historians have been investigating knowledge.

Starting with the idea that good ignorance springs from knowledge, we might begin by looking at some of the limits on knowledge in science and see what their effects have been on ignorance generation, that is, on progress.

THREE

Limits, Uncertainty, Impossibility, and Other Minor Problems

The notion of discovery as uncovering or revealing is in essence a Platonic view that the world already exists out there and eventually we will, or could, know all about it. The tree falling in an uninhabited forest indeed makes noise—as long as noise is defined as a simple physical process in which air molecules are forced to move in compression waves. That they are perceived by us as "sound" simply means that evolution hit upon the possibility of detecting this movement of air with some specialized sensors that eventually became our ears. Now, of course, there are things going on out there that evolution perhaps ignored—leading to our ignorance of them. For example, consider the wide stretches of the electromagnetic spectrum, including most obviously the ultraviolet and infrared but also several million additional wavelengths that we now detect only by using devices such as televisions, cell phones, and radios. All were completely unknown, indeed inconceivable, to our ancestors of just a few generations ago.

It is a fairly clear and simple point to understand that our sensory apparatus, molded by evolution to enable us to find food for ourselves and avoid becoming food for someone else long enough to have sex and produce offspring, is not capable of perceiving great parts of the universe around us. But that same evolutionary process molded our mental apparatus as well. Are there things beyond its comprehension? Just as there are forces beyond the perception of our sensory apparatus, there may be perspectives that are beyond the conception of our mental apparatus. The renowned early 20th-century biologist J. B. S. Haldane, known for his keen and precise insights, admonished that "not only is the universe queerer than we suppose, it is queerer than we *can* suppose." Since then we have discovered neutrinos and quarks of various flavors, possible new dimensions, long molecules of a snotty substance called DNA that contains our genes, antibodies that recognize us from others, and we have used this and other knowledge to invent television, telecommunications, and an endless list of truly amazing things. And for all of this, Haldane's aphorism actually seems more correct and relevant now than when he uttered it in 1927.

In a similar vein, Nicholas Rescher, a philosopher and historian of science, has coined the term *Copernican cognitivism*. If the original Copernican revolution showed us that there is nothing privileged about our position in space, perhaps there is also nothing privileged about our cognitive landscape either. In Edwin Abbott's 19th-century fantasy novel, a civilization called Flatland is populated by geometric beings (squares, circles, triangles) that live in only two dimensions and cannot imagine a third dimension. It is surprisingly easy to identify with the lives of these creatures, leaving one to wonder whether we don't all live in a place that is at least one dimension short. The inhabitants of Flatland are mystified and terrified by the appearance one day of a circle that can magically change its circumference. It appears from nowhere as a point, grows slowly to a small circle, becomes larger and larger, and then just as smoothly diminishes in size until it is a point again, and then, incomprehensibly to the Flatlanders, disappears. It is of course just the observation of a three-dimensional sphere passing through the two-dimensional plane of Flatland. But this simple solution is inconceivable to the inhabitants of Flatland, just as it is almost inconceivable to us that they could be so stupid, no matter that the 11 or so dimensions proposed by string theory are well beyond our conception (or the physical limits of our senses).

Let's take an example from the history of science. Since the Greeks started it all, there has been ongoing controversy in science as to whether the world is composed essentially of a very large number of very small particles (atomism) or is a continuum, a wave not a particle, a smooth progression of time only falsely and arbitrarily divided up into seconds or minutes, a single expanse of space not divided by degrees and coordinates. As Bertrand Russell is claimed to have remarked, is the universe a bucket of sand or a pail of molasses? We tend to see the continuum better than the discrete entities because the infinitesimal is not available to our senses. Is this what stands in the way of our breaking through the apparent paradoxes of quantum physics? Is it a shortcoming in our perceptual and cognitive apparatus?

There is a kind of discomfort that arises from this line of reasoning. As if there were things going on, right under our noses, that we didn't know about. Worse than that, *couldn't* know about. And even more discomforting is that we may never have the capability to know about them. There may be limits. If there are sensory stimuli beyond our perception, why not ideas beyond our conception? Have we run into any of those limits yet? Would we know them if we did? Comedian philosopher George Carlin wryly observed that "One can never know for sure what a deserted area looks like."

OFFICIAL LIMITS

In science there are so far two well-known instances where knowledge is shown to have limits. The famous physicist Werner Heisenberg's Uncertainty Principle tells us that in the subatomic universe there are theoretically imposed limits to knowledge—the position and momentum of a subatomic particle (as well as other pairs of observations) can never be known simultaneously. Similarly, in mathematics, Kurt Gödel in his Incompleteness Theorems demonstrated that every logical system that is complex enough to be interesting must remain incomplete. Are there other limits like these? For example, in biology some ask whether a brain can understand itself. Turbulence or the weather may be fundamentally unpredictable in ways we don't yet grasp. We don't know. Do they matter? Surprisingly they don't really have much effect on vast parts of the scientific enterprise, at least not as much as some metaphysically minded writers would have you believe. Why not? Let's have a closer look—for those among us acutely aware of our ignorance, it is sometimes instructive to see how not knowing may not matter.

In the sphere of subatomic particles, "uncertainty" does make a difference, but this is a very rarefied place, and generally of little concern to corporeal beings such as ourselves. But it is a useful example of a limitation that rose up unexpectedly and could have set physics on its ear. In fact, it revealed new and previously unknown unknowns, it gave rise to decades of fruitful and unanticipated advances, and it created stranger and yet more interesting problems that remain active areas of inquiry today. Entanglement, one of the most peculiar results in the whole mad zoo of quantum physics, grew almost directly from the required uncertainty unveiled by Heisenberg.

Heisenberg's result is not simply a case of lacking a good-enough measuring device. The very nature of the universe, what is called the wave-particle duality of subatomic entities, makes these measurements impossible, and their impossibility proves the validity of this deep view of the universe. Some fundamental things can never be known with certainty. And the hard fact is that if you can't measure starting values, you can never predict the future state. If you can't measure the position (or the momentum) of a particle at time zero, you can't know, for sure, where the particle will be at any future time. The universe is not deterministic; it is probabilistic, and the future can't be predicted with certainty. Now it is true that, as a practical matter, for things that have masses greater than about 10^{-28} grams, the probabilities become so large that predicting how they will act is quite possible—baseball players regularly predict the path of 150 gram spheres on a trajectory covering 100 meters and if a shoe that has been thrown by an irate journalist is coming at your head from the right, ducking to the left is certainly a good bet. Unfortunately it is just this discontinuity in scale between the quantum and the inhabited worlds that makes quantum uncertainty so difficult to appreciate. As many of the pioneers in quantum mechanics noted, these phenomena can only be understood by willingly forgoing any sensible (i.e., sensory-based) description of the world. How ironic that the weird but undeniable results of quantum mechanics rest on a rigorous mathematical scaffold, even while it is conceptually available only in metaphorical allusions like "entanglement" or Schrodinger's cat that is at once alive and dead and neither. But regardless of whether you can grasp it, the important thing to know about quantum uncertainty is that, whatever it may look like, it actually has not been a limitation; rather it has actually spawned more research, more inquiry, and more new ideas. Sometimes limitations on knowledge can be very useful.

Then there is Gödel's daring challenge to the completeness of mathematics. The diminutive and unassuming Gödel began hatching his ideas at a time when scientific and philosophical thinking was dominated by positivism, the oversized and intellectually aggressive belief that everything could be explained by empirical observation and logic because the universe, and all it contained, was essentially mechanistic. In mathematics this view was advanced especially by David Hilbert, who proposed a philosophy called formalism that sought to describe all of mathematics in a set of formal rules, axioms to a mathematician, that were logical and consistent and, well, complete.

He was not the first or the only great mathematician to have this dream. The 17th-century German philosopher and mathematician Gottfried Leibniz, one of the inventors of calculus, had a lifelong project to construct a "basic alphabet of human thoughts" that would allow one to take combinations of simple thoughts and form any complex idea, just as a limited number of words can be combined endlessly to form any sentence—including sentences never before heard or spoken. Thus, with a few primary simple thoughts and the rules of combination one could generate computationally (although in Leibniz's day it would have been mechanically) all the possible human thoughts. It was Leibniz's idea that this procedure would allow one to determine immediately if a thought were true or valuable or interesting in much the same way these judgments can be made about a sentence or an equation—is it properly formed, does it make sense, is it interesting? He was famously quoted as saying that any dispute could be settled by calculating—"Let us calculate!" he was apparently known to blurt out in the middle of a bar brawl. It was this obsession that led Leibniz to develop the branch of mathematics known today as combinatorics. This in turn sprang from the original insight that all truths can be deduced from a smaller number of primary or primitive statements, which could be made no simpler, and that mathematical operations (multiplication was the one Leibniz proposed but also prime factorization) could derive all subsequent thoughts. In many ways this was the beginning of modern logic; indeed, some consider his *On the Art of Combinations* the major step leading from Aristotle to modern logic, although Leibniz himself never made such claims. Does it somehow seem naive for him to have proposed that we could think of everything if we just built this little calculating device and put in a few simple ideas? Leibniz himself seems to have recognized his naivety as he notes that the idea, which came to him when he was 18 years old, excited him greatly "assuredly because of youthful delight." Nonetheless, he was obsessed with this "thought alphabet" and its implications for most of the rest of his life, and the *On the Art of Combinations*, which was part of the project, introduced a powerful new mathematics.

The interest for us, and possibly for Leibniz, is not simply that this structure, this imaginary device, could construct all human thoughts, but that it could also identify and appraise thoughts that are unknown. It could survey not only what we know but also what we do not know. It is this attribute that holds the enticement, and indeed it may contain more of what we do not know—just as there are likely to be more sentences still unuttered than all those that have been spoken. The power of Leibniz's thought alphabet and the thought algebra that would run it was not how well it would settle disputes but that it showed the infinity of human thought, the immensity of the unknown. Language is useful for what it allows one to say, but it is powerful because it admits, by its very structure, that there are an infinity of things that could be said, and that there will always be more unsaid than said. That the Leibniz alphabet never came to be used the way that his youthful vision had imagined is less important than the demonstration that simple things can be combined to make endless new compound things. We are now developing whole new branches of science to analyze and manage and manipulate this complexity.

Two hundred years later Hilbert's positivism program was another attempt to codify knowledge, but this one was doomed by the seemingly simple ruminations of Kurt Gödel, another German mathematician who had become interested in logic. As Rebecca Goldstein recounts in her excellent, and highly detailed, book on Gödel, his shyness and reluctance to make big pronouncements, perhaps the opposite of Hilbert's style, at first obscured the explosive power of his incompleteness theorems. Eventually, however, it dawned on the mathematics community: There would be no axiomatic, consistent, *and* complete theory of mathematics. A consistent system in mathematics, such as the familiar whole-number arithmetic and its operations (addition, subtraction, etc.), can never be complete as well as consistent. Consistency refers to the straightforward characteristic that a system's rules will not result in contradictory statements, for example, that two things are and are not equal. Although this may seem simple, it is devilishly difficult to be sure that a few apparently simple statements do not, in any of the myriad possible ways they may be used (combinatorially, Leibniz would say), ever lead to an illogical conclusion. Introducing seemingly reasonable concepts like "zero" or "infinity" into simple arithmetic, for example, can result in strange incompatibilities (*antinomies,* mathematicians call them). The challenge is to show that for a particular system the root axioms, the basic fundamental rules, will never result in such incompatibilities. Proving this means that the system is *complete*. What Gödel showed, using a strange new correspondence between mathematics and logic that he invented, was that if a system were consistent it could never be shown to be complete within the rules of that system. This means that something that could be shown to be true using the system could not in fact be proved to be so. Since proofs are the foundation of mathematics, it is quite curious when obviously true statements cannot be proved. The math is complicated beyond the scope of this book, but the gist of it can be appreciated by considering any of several paradoxes that bend your brain in unpleasant ways. The most famous of these is the Cretin's paradox, sometimes known as the Liar's paradox. They all go something like: "The Cretin claims that all Cretins are liars." So who are you to believe? Or another version—take a blank card and write on one side of it, "The statement on the other side of this card is true" and on the other side write, "The statement on the other side of this card is false." These little mind games, became for Gödel, the basis of a new form of logic that he used to demonstrate that in many circumstances you can't tell yourself the truth.

Was this the end of the messianic program to establish the primacy of mathematics and of logical thinking? As it turns out, quite the contrary. Gödel's small, by comparison, but revolutionary output is so astonishing because of the technical and philosophical research opportunities it has created. Previously unconsidered ideas about recursiveness, paradox, algorithms, and even consciousness owe their foundations to Gödel's ideas about incompleteness. What at first seems like a negative—eternal incompleteness—turns out to be fruitful beyond imagining. Perhaps paradoxically much of computer science, an area one might think was most dependent on empirical statements of unimpeachable logic, could not have progressed without the seminal ideas of Gödel. Indeed, unknowability and incompleteness are the best things that ever happened to science.

So some things can never be known and, get this, it doesn't matter. We cannot know the exact value of pi. That has little practical effect on doing geometry. As Princeton astrophysicist Piet Hut points out, the early Pythagoreans were stopped for a while in their tracks when they realized that the square root of 2 could not be precisely represented on the number line, the continuum that translated the numbers of counting into smooth distances. You cannot cut the line at the point corresponding to the $\sqrt{2}$ and have two new lines that add up to the old one. Very disturbing if the value for the length of the hypotenuse of the simplest right triangle, one with each of its sides equal to 1, does not have a particular location anywhere on the number line from minus to plus infinity. Yet there is a very strong proof of this apparent paradox. A traditional, although possibly apocryphal, story has it that one of the Pythagoreans, Hippasus, upon showing his proof of this strange and, at the time, heretical finding, was drowned by his fellow Pythagoreans. This was a nasty consequence for getting the right answer; math, it seems, was much tougher in those days. But after a time mathematicians developed a work-around. It turns out there are other numbers like $\sqrt{2}$, and they are called irrational numbers—not because they are unreasonable but because they cannot be expressed as a fraction, that is, as a ratio of two other numbers. The irrational numbers, along with the more common rational numbers that do have spots on the number line, make up what we now call the set of *real* numbers. Now we can work with them more or less as we would with rational ("normal") numbers and no one worries about it anymore. You don't, do you? Probably never even occurred to you.

. . .

We now have an important insight. It is that the problem of the unknowable, even the really unknowable, may not be a serious obstacle. The unknowable may itself become a fact. It can serve as a portal to deeper understanding. Most important, it certainly has not interfered with the production of ignorance and therefore of the scientific program. Rather, the very notions of incompleteness or uncertainty should be taken as the herald of science.

This leads to a second insight regarding ignorance. If ignorance, even more than data, is what propels science, then it requires the same degree of care and thought that one accords data. Whatever it may look like from outside the science establishment, the incorrect management of ignorance has far more serious consequences than screwing up with the data. There are correction procedures for mishandled data—they must be replicable, must answer to the scrutiny of peers—but mishandled ignorance can be costly, harder to perceive, and so harder to correct.

When ignorance is managed incorrectly or thoughtlessly, it can be limiting rather than liberating.

Scientists use ignorance to program their work, to identify what should be done, what the next steps are, where they should concentrate their energies. Of course, there is nothing wrong in principle with laying out what you need to know—this is what grant proposals are supposed to accomplish. But as any working scientist will tell you, what gets proposed in the grant and what gets done during the actual period of grant funding are not often very similar. I speak from experience, but it is a common one. Things happen, or don't, that redirect your thinking; work from other laboratories reveals a new result that requires you to revise your ideas; results from your own experiments are not what you expected and force new interpretations and new strategies. The goals may remain similar, but the path changes because the ignorance shifts. Thomas Huxley once bemoaned the great tragedy of science as the slaying of a beautiful hypothesis by an ugly fact—but nothing is more important to recognize, either. Grieve and move on.

Ignorance then is really about the future; it is a best guess about where we should be digging for data. Can we learn something about managing ignorance from how these guesses are made? How does this view of the future direct scientific thinking?

FOUR

Unpredicting

> Who among us would not be happy to lift the veil behind which is hidden the future; to gaze at the coming developments of our science and at the secrets of its development in the centuries to come?
>
> —David Hilbert, introduction to his speech at the Second International Congress of Mathematicians held in Paris, 1900

> The future ain't what it used to be.
>
> —Yogi Berra, American philosopher, baseball player, and team manager

Predictions come in two flavors in science. One is about the direction of future science. The other one, equally if not more important to the everyday mechanics of science, is the ability of science to make testable predictions. An experiment is designed to test the most general principle possible, even though it is almost always only a particular instance of that principle. Thus, a chemist wants to test the validity of a reaction between two elements under certain conditions and designs an experiment in which these two elements are brought together and the result of their interaction can be measured—how much heat is put out, what new molecules have appeared, how much of the original material is left, and so forth. By doing this he or she hopes to come up with a general rule about this type of reaction, such that over a wide range of particulars (the amount of stuff you start with, the initial conditions, etc.) anyone can predict the outcome. If an outcome can be reliably predicted from a limited amount of starting information, then you have gained an understanding of an underlying principle, of the rules governing this bit of the universe. A particular set of genes predicts the likely color of your hair or eyes; two massive bodies at a certain distance will orbit each other with a particular period. These are all instances where knowing the underlying mechanism allows you to make reliable predictions about outcomes. In science, predicting is knowing.

As I no longer need to tell you, this book is very specifically not about knowing, so that's why I'm going to concentrate on the other side of predicting in science. By this I mean the sort of predicting Hilbert had in mind when he opened the Congress of Mathematicians in 1900 with the statement that opened this chapter: to see where science will take us, what new mysteries it will present, to imagine the future.

Predicting the coming advances in science and technology is a common if often silly exercise, mostly the provenance of magazine editors who see it as a requirement for their end of the year, decade, or millennium issues. Scientists are interviewed and asked what they see as the likely advances in their fields over the next decade or so. Being a generally optimistic lot, at least in their public face, they tend to tackle questions of this sort with gusto, invariably leading to inflated prognostications from the fantasy wish list that every scientist has tucked away in a desk drawer. Unbridled enthusiasm for scientific progress is good public relations, but it is often bad science. Things never go the way we think they will; there are always unexpected findings and unexpected consequences that may redirect or even stymie a field for years.

In fact, one of the most predictable things about predictions is how often they're wrong. Nonetheless, they are a measure, even if somewhat imprecise, of our ignorance. They are a catalog of what we think the important ignorance is, and perhaps also a judgment of what we think is the most solvable ignorance. David Hilbert was probably the most successful at this game. In the talk that followed that opening comment in August 1900, he outlined 23 crucial problems for mathematics to solve in the next century. These problems, now known eponymously as the Hilbert problems, dominated mathematical research throughout the 20th century. Hilbert was a successful prognosticator because he cleverly turned the tables: his predictions were questions. His predictions were truly a catalog of ignorance because they simply set out what was unknown and suggested that this is where mathematicians might be wise to spend their time. The result is that slightly more than a century later 10 of the 23 problems have been solved to the satisfaction of a consensus, the others being partially solved, unsolved, or now considered unsolvable.

So Hilbert's strategy, one that we might do well to learn from, was to predict ignorance and not answers. He put no time line on when the major problems might be solved, nor even if they would be solved, but nonetheless there are few mathematicians who would not agree that Hilbert's little speech at the opening of the 20th century was a positive influence on mathematics that effectively set much of the field's agenda for more than a hundred years.

When used this way, predicting scientific progress becomes more than just an exercise because it finds its way into making science policy, where it can have either positive or negative effects on determining how limited resources are spent on research. This is why it is as important to be careful with ignorance, no less so than with the facts. Granted, it is reassuring, when budgeting billions for scientific research, to believe that there is a rational program that can be mapped and followed to produce some set of desired results, or at least something that can be called progress. But this is a false assurance based on unreliable judgments about ignorance. It's hard to see what will be and also what will not be. We are not flying about with individual jet packs, we are not wearing disposable clothes or dining on concentrated nutrients in foil packs, and we have not eradicated malaria or cancer, all predicted years ago as likely. But we do have an Internet that connects the entire world, and we do have a pill that provides erections on demand—neither of which will be found in any set of published predictions from 50, or even 25 years ago. As Enrico Fermi noted, predictions are a risky business, especially when they are about the future.

So how should our scientific goals be set? By thinking about ignorance and how to make it *grow*, not shrink—in other words, by moving the horizon. Predicting or targeting some specific advance is less useful than aiming for deeper understanding. Now this may sound like just so much screwing around, but again and again this is how most of the great advances in science and technology have occurred. We dig deeper into fundamental mechanisms and only then does it become clear how to make the applications. Whether it is lasers, X-rays, magnetic resonance imaging (MRI), or antibiotics, applications are surprisingly obvious, once you understand the fundamentals. They are just shots in the dark if you don't.

Let's take an example. In 1928 the eminent physicist Paul Dirac was trying to describe the electron in quantum mechanical terms. He derived what has become known as the Dirac equation, a rather complex mathematical formulation that neither you (unless you are a trained physicist) nor I can understand. What we can understand is that while the equation filled in some fundamental gaps in nuclear theory, it also raised many serious new questions—some of which are still around. One of those questions was that the equation predicted an anti-electron, a particle with all the electron's properties but of opposite charge—a positron. No one had ever seen this particle in any experiment, and Dirac himself expressed some doubts about ever observing such a particle, but according to his calculations, which explained an awful lot, it had to be there. It was this glimpse of ignorance that led to new experiments, and in 1932, using a technology known as "cloud chambers" (later, "bubble chambers"), physicist Carl Anderson observed the track created in his chamber by a positron, thereby discovering what Dirac had predicted 4 years earlier. If you had asked Dirac or Anderson what the possible applications of their studies were, they would surely have said their research was aimed simply at understanding the fundamental nature of matter and energy in the universe and that applications were unlikely, and certainly outside of their interest. Nonetheless in the late 1970s biophysicists and engineers developed the first PET scanner—that stands for *positron* emission tomography. Yes, *that* positron. Some 40 years after Dirac and Anderson, the positron came to be used in one of the most important diagnostic and research instruments in modern medicine. Of course, a great deal of additional research went into this as well, but only part of it was directed specifically at making this machine. Methods of tomography, an imaging technique, some new chemistry to prepare solutions that would produce positrons, and advances in computer technology and programming—all of these led in the most indirect and fundamentally unpredictable way to the PET scanner at your local hospital. The point is that this purpose could never have been imagined even by as clever a fellow as Paul Dirac.

The problem with the dichotomy between basic and applied research is that it is fundamentally false—that's why it never seems to get solved and we endlessly oscillate back and forth in favor of one, then the other—as if they were two things and not just one research effort. Following the ignorance often leads to marvelous inventions. But trying to take short cuts, to short circuit the process by going directly to the application, rarely produces anything of value. Thus, for example, the vast amount of work that has been expended on trying to make computers converse as if this was just a programming problem and not a deep issue of cognitive neuroscience. Where, finally we must ask ourselves, is the source of the inventions—is it from Edisons or Einsteins? Given the choice, which would we want more of, Edisons or Einsteins? Edison was a great inventor, but without the understanding of electricity that came from Faraday's basic experiments and mathematical formulations, he could have done none of it, wouldn't have even thought of doing any of it. Granted, it often takes an Edison to make something of Faraday or Einstein's pure knowledge, but carts before horses don't go anywhere. Faraday, by the way, had no idea what electricity might be good for and responded to a question about the possible use of electromagnetic fields with the retort, "Of what use is a newborn baby?" This phrase he apparently borrowed from Benjamin Franklin, no less, who was the first to make the analogy in his response to someone asking him what good flight would ever be after witnessing the first demonstration of hot air balloons. People who want to know what use something is rarely seem to have much imagination.

One favorite style of the predictions issues of magazines is to number them—the "50 Greatest Advances for the next 50 years" or the "10 Greatest Enigmas in Science." This is a subtly dangerous approach as well. I am sure that those who engineer these articles mean no harm, but enumerating ignorance in this way is to make us believe that we can see the horizon, that we can get there, that it will not infinitely recede, and that there is a finite number of scientific problems to solve and then that will be that and we can get on with the leisurely utopian part of humanity's saga. Numbering, in this case, places limits where there are none and at its worst drives us to direct science at false goals that are often unattainable and are ultimately a waste of money and other resources. Numbering leads to prioritizing—the accounting alternative to creativity.

There is also a certain conclusive, but wrong, notion that comes from an explicit number. In a peculiar way it is an ending, not a beginning. A recipe to finish, not to continue. One might say that Hilbert's "23 problems" suffers a bit from this, but perhaps it's just in the nature of mathematicians to number things, including even their ignorance. For the rest of science it seems wiser not to enumerate so precisely but to take the important lesson that predicting ignorance, not accomplishments, is more fruitful—and less likely to be wrong.

Ignorance works as the engine of science because it is virtually unbounded, and that makes science much more expansive. This is not just a plea for unlimited science; it may well all come to an end one day for any of a variety of reasons, from economic to social to intellectual. Rather it is an argument for the view that as long as we are doing science it is better to see it as unbounded in all directions so that discovery can proceed everywhere. It is best not to be too judgmental about progress.

However, this doesn't mean we should just go off in whatever direction our whims take us and hope for the best. Ignorance is not just an excuse for poor planning. We must think about how ignorance works, and we have to be explicit about how to make it work to our advantage.

While for many experienced scientists this is intuitive, it is not so obvious to the layperson, and it often seems not so apparent to young scientists starting out their career and worrying about grant support and tenure. Let me take a stab at analyzing ignorance more deeply.

FIVE

The Quality of Ignorance

We can see from these so far straightforward arguments that ignorance is not so simple a concept. In its less pejorative uses it describes a productive state of scholarship, experimentation, and hypothesizing. It is both the beginning of the scientific process—and its result. It is the beginning, of course, because it asks the question. "It is always advisable to perceive clearly our ignorance," said Charles Darwin early in his book *The Expression of the Emotions in Man and Animals*. The ignorance of a subject is the motivating force. At first, it is most of what we know. Insufficiently considered ignorance is problematic. Just saying we don't know something is not critical or thoughtful enough. It can lead to questions that are too big, or too amorphous, or too hard to imagine solving. Thoroughly conscious ignorance is, as Maxwell had it, the prelude to discovery.

It is the product of science as well. Although not the explicit goal, the best science can really be seen as refining ignorance. Scientists, especially young ones, can get too enamored with results. Society helps them along in this mad chase. Big discoveries are covered in the press, show up on the University's home page, garner awards, help get grants, and make the case for promotions and tenure. But it's wrong. Great scientists, the pioneers that we admire, are not concerned with results but with the next questions. The eminent physicist Enrico Fermi told his students that an experiment that successfully proves a hypothesis is a measurement; one that doesn't is a discovery. A discovery, an uncovering—of new ignorance.

The Nobel Prize, the pinnacle of scientific accomplishment, is awarded, not for a lifetime of scientific achievement, but for a single discovery, a result. Even the Noble committee realizes in some way that this is not really in the scientific spirit, and their award citations commonly honor the discovery for having "opened a field up," "transformed a field," or "taken a field in new and unexpected directions." All of which means that the discovery created more, and better, ignorance. David Gross, in his acceptance speech for the Nobel Prize in Physics (2004), noted that the two requirements for continuing Nobel Prizes were money, kindly supplied by Alfred Nobel's bequest, and ignorance, currently being well supplied by scientists.

Okay, so we're convinced that ignorance is worth taking seriously. But how do scientists actually work with ignorance; specifically, how does it show up in their day-to-day work in the lab or the way they organize their labs and plan experiments? The first thing to recognize is that ignorance, like many such big meaningful words, fails to describe the breadth of its subject—or rather it describes only the breadth, missing the many details within its depths. Ignorance comes in many flavors, and there are correspondingly many ways of working with it. There is low-quality ignorance and high-quality ignorance. Scientists argue about this all the time. Sometimes these arguments are called grant proposals; sometimes bull sessions. They are always serious. Decisions about ignorance may be the most critical ones a scientist makes.

Perhaps the first thing for a scientist to consider is how to decide, against the enormous backdrop of the unknown, what particular part of the darkness he or she will inhabit. My laboratory works on olfaction, the sense of smell. It is a small subfield within the larger field of sensory systems that include vision, hearing, touch, taste, and pain. "Sensory systems" is itself a subfield within the much larger discipline of neurobiology, the study of nervous systems. And that, in turn, is just one area of investigation within the still larger domain known as biology, itself encompassing ecology, evolution, genetics, physiology, anatomy, zoology, botany, biochemistry, and so on. The Society for Neuroscience, the professional society representing workers in all fields of neuroscience, boasts a membership of over 40,000 and holds an annual meeting attended by more than 30,000 scientists and educators. How do all these scientists sort themselves out? Why don't they all work on the same thing or one of a few (e.g., 23) things—memory, schizophrenia, paralysis, stroke, or development? Aren't these the big questions in neuroscience? Aren't these the hot topics you watch presented in slick documentaries on public or cable television?

How do scientists, as opposed to TV producers, ponder these big questions about ignorance? How do they get from these and other interesting and important issues to an actual scientific research program? Well, at the most pedestrian, but nonetheless critical level, there are grant proposals. Every scientist spends a significant percentage of his or her time writing grants. Many complain about this, but I actually think it's a good idea. These documents are, after all, a detailed statement of what the scientist hopes to know, but doesn't, as well as a rudimentary plan for finding it out. Scientists write grant proposals that are reviewed by other scientists, serving unpaid on government committees, who recommend what they consider the best of the proposals for funding. These grant proposals, numbering many thousands per year, represent a virtual marketplace of ignorance. Imagine being awarded a prize for what you don't know: Here's some money for what you don't know. Everyone else in the world is getting paid for what they know—or claim to know. But scientists get rewarded for their ignorance. If that's the case, then it can't just be for any old ignorance. It has to be really good ignorance. One must become an expert, a kind of connoisseur of ignorance. In its most unkind characterization this might be called grantsmanship. But this is unfair. The art of writing a grant, of writing about ignorance authoritatively, is not trivial.

How does one develop into a connoisseur of ignorance? There are numerous strategies, and I will endeavor to list and describe some of them in the discussion that follows. To be honest, though, it is often a matter of intuition and taste. As you will see, questions can be tractable or intractable, interesting or ordinary, narrow or broad, focused or diffuse—and any of all the possible combinations of those attributes. There is not a single Method of Ignorance. While I would like to provide a simple-to-follow Handbook of Ignorance, I cannot in fact be prescriptive. One of the surprising things that I learned from teaching a class on ignorance is that science is remarkably idiosyncratic. Individual scientists, although bound together by a few crucial rules about what will pass muster, otherwise take quite distinctive approaches to how they do their work. So what

I present to you here is a Potpourri of Ignorance, a Multiplicity of Ignorance. It will sometimes seem conflicted, one strategy will be at odds with the following and preceding ones, but that's actually the way it is. There are many strategies of ignorance. I have come to appreciate this richness, but I understand that it may be bewildering at first. Bear with me.

THE MANY MANIFESTATIONS OF IGNORANCE

Let's begin with what makes a question interesting? Mathematicians often use this term when they say that so and so a conjecture is correct but not interesting. When I ask Columbia University mathematician Maria Chudnovsky, who works in a very specialized area called Perfect Graph Theory (which by the way has nothing to do with the graphs you and I are familiar with), she says that a question is interesting if it leads somewhere and is connected to other questions. Something can be unknown, and you test it out for a bit, but then you can see, often pretty quickly, that it is not very connected to other things that are unknown and therefore it is not likely to be interesting or worthy of pursuit. If it seems as though you are working away on a project and nothing that anyone else is doing or has done becomes helpful to your work, then you begin to think that you are perhaps in some cul-de-sac of irrelevance. This happens to graduate students quite often. They start a project with a question that is mostly untouched or has received little attention. But some ways into it, the data don't seem to lead anywhere, they keep proving the same small thing over and over again, and eventually there is nothing to do but abandon the project. So connectedness seems to be an important quality.

On the other hand, biology is full of people working on a barely known species of organism, from a virus to a mammal, that has some quirky lifestyle and that they find immensely interesting, perhaps because it is not in any obvious way connected to the mainstream of biology. Sometimes these apparent cul-de-sacs become part of the mainstream in very unexpected ways, suddenly connecting up to the main branch and bringing new comprehension to questions that no one had even thought about asking. Just as often they remain dead ends. But like Darwin and his worms, the biologist's curiosity is enough for him or her to spend a lifetime mastering the details of another creature's life history. This kind of work takes a certain faith that it will all mean something someday. Or maybe it just takes a laissez-faire attitude that not everything has to mean something.

One example of curiosity-driven research that unpredictably produced one of the crucial tools in the biotechnology revolution is the study of *thermophiles*, a word that literally means "heat loving." What a wonderful word.

I can't help thinking of it when I walk along a beach in southern Florida watching hordes of people lying in the radiation, risking melanoma and wrinkles, and loving every minute of it. But nature's real thermophiles thrive in the inferno of near-boiling, sulfurous, deep sea vents and in the hot sulfur springs of Yellowstone National Park. This is where they were first discovered in the 1960s by microbiologist Thomas Brock of Indiana University and an undergraduate in his laboratory named Hudson Freeze (I'm not making that up). These microorganisms, at first an oddity, became suddenly important because their enzymes had adapted to the high temperatures of their niche, temperatures that would cause similar enzymes in our bodies to disintegrate. Then 30 years later, in the 1990s, temperature-resistant enzymes were precisely what was required for reactions like those in polymerase chain reaction, known more commonly as PCR, the technique that is fundamental to most biotechnology experiments and ubiquitous on forensic crime shows. PCR works by cycling through temperatures that vary from 40°C to 98°C (approximate body temperature to nearly boiling) and requires enzymes that can withstand, and function, at these high temperatures. Once again, apparently isolated research, undertaken only for the sake of curiosity, comes to play a critical but completely unpredicted role in the discovery and invention of new technologies and products.

Here are some other ways that scientists think about ignorance, presented in no particular order, because no particular order presents itself. Or because, to tell you the truth, there is no particular order. They are all more or less equivalent.

One way is through the lens of what the Nobel laureate immunologist Peter Medawar called "The Art of the Soluble." Medawar claims that simply showing something is possible can be sufficient to motivate work and make progress. His Noble Prize–winning work was to show that the immune system can recognize self from other in all its tissues and how this occurs. By explaining the biological basis for the well-known phenomenon of organ rejection, Medawar is often credited with making organ transplantation possible. Medawar, however, eschews this, saying that all he did was to show that in principle it was not impossible—all that was needed was to find a way to fool the immune system into accepting "other" as self. So just showing something is solvable is one strategy. What sorts of ignorance can we erase? What questions look like they could be answered? After all, there is no sense in banging your head against the wall; why not apply yourself to something tractable?

A story many of us tell our graduate students is about a scientist searching the ground under a street lamp late at night. A man walks up to him and asks him what he has lost. "My car keys," says the scientist, and the friendly chap helps him look. After a while, and no success, the fellow asks whether he is sure this is where he dropped them. "No, I think probably over there," he says pointing to a dark portion of the street. "Then why are you looking here?" "Well," says the canny scientist, "the light is so much better here." This story is often told in a way to make the seeker seem ridiculous (in fact, in some tellings he is not a scientist but a drunk, or maybe a drunken scientist), but I think it's just the opposite. One very decent strategy in science is to look where you have a good chance of finding something, anything. The lesson here is to recognize the value of the observable and to leave the unmeasurable stuff for later. Anyway, if you're drunk it's better not to find your car keys.

. . .

An almost opposite strategy can be summed up by the parable that began this book: *it is very difficult to find a black cat in a dark room—especially when there is no cat.* This is ignorance driven by deep mysteries. One enters the room and stumbles around, the black cat is reported to be in here, but no one has seen it directly, and the reports are of questionable reliability. In science there are dark rooms everywhere that have been found to be completely empty, each one representing careers that in whole or part have been spent finding out this important but not very satisfying fact. False leads are followed, seemingly good ideas and reasonable theories are pursued, only to find out that they were piti-fully mistaken, fundamentally incorrect. This is the fear of every scientist. But it is also the motivation, the excitement, the thing that gets you to the lab early and keeps you there until late at night. It is also the part of science that most nonscientists miss entirely.

It gets missed because we rely so heavily on the newspaper or the TV for information about what's going on in science, where only the black cats that get discovered are featured. We rarely hear of the pursuits, especially the unsuccessful or not-yet-successful ones. The reports from the frontier are unfortunately "improved" by highlighting the findings and ignoring the process—ignoring the ignorance, if you will.

I say this is unfortunate because it has two unwanted effects. First, it makes science seem inaccessible, because how could you ever keep up with the steady stream of new facts (remember the 5 exabytes of new information in 2002, the 1 million new scientific publications last year). Second, it gives a false impression of science as a surefire, tough-as-nails, deliberate method for finding things out and getting things done, when in fact the process is actually more frag-ile than you might imagine and requires more nurturing and more patience (and more money) than we commonly think is the case. Einstein, responding to a question about why modern science seemed to flower in the West rather than India or China (at that particular time in history this was largely the case), remarked that what was puzzling was that it was here at all, not why it wasn't in India or China. Science is a risky business. For some scientists that is a reason to stay with the more tractable questions; for others the risk seems to be what makes it worthwhile.

Again, though, there is the flip side. Faced with black cats that may or may not be there, some scientists are happy instead to measure the room—its size, its temperature, its age, its material composition, its location—somehow forgetting about or ignoring the cat. Perhaps this has a ring of timidity to the reader, of a concern with the mundane rather than the extraordinary, but in fact measurement is critical to advancing science. Much that is good and valuable has come from just this sort of quotidian scientific activity. Many of the comforts of modern life, not to mention the amelioration of many miseries suffered by our ancestors, have come from the work of scientists who make these measurements. Kepler spent 6 years battling with an error of 8 minutes of arc in the planetary motion of Mars (that's an amount of sky about equal to one-third the width of your thumb held at arm's length). But the result of this attention to measurement and exactitude was that he freed astronomy from the Platonic tyranny of the perfect circle and showed that planets move around the sun in ellipses. Newton could never have understood motion and gravity if he hadn't had this critical advance in front of him.

Advances in measurement techniques almost always precede important new discoveries. Facts that seem settled at 5 decimal places become ambiguous at 6 or more. The desire to measure more accurately drives technology and innovation, resulting in new microscopes with more resolution, new colliders with more smashing power, new detectors with more capturing capability, new telescopes with more reach. And each of these advances in turn makes searching for black cats more tractable. Ignorance of the next decimal place is a scientific frontier no less grand than theorizing about the nature of consciousness or some other "big" question.

. . .

The ignorance in one's own professional backyard is sometimes the most difficult to identify. The journals *Nature* and *Science* are published weekly and contain reports that are judged to be of especially high significance. Getting a paper in one of these journals is the science version of landing a leading role or winning a big account. For many it can make a career, or at least get one started on the right foot. Each week doctoral and postdoctoral students in labs around the world scour the pages of these journals for the latest finding in their field and then try to think of the next experiment so that they can get to work on their *Nature* paper. But of course it's already too late; the folks who wrote that paper have already figured out the next experiments—in fact, they've probably just about finished them. I have a colleague who always suggests that his students look not to yesterday's issue of *Nature* or *Science* for experimental ideas but rather to work that is at least 10 or more years old. This is work that is ready to be revisited, ready for revision. Questions still lurk in these data, questions that have now ripened and matured, that could not be answered then with the available techniques. More than likely they could not even have been asked because they didn't fit any current thinking. But now they come alive, suddenly possible, potential, promising. Here is another fertile, if unintuitive, place to look for ignorance—among what's known.

. . .

How big should a question be? How important should it be? How can you estimate the size or importance of a question? Does size matter? (Sorry, how could I resist?) There are no answers to these questions, but they are nonetheless good questions because they provide a way to think about...questions. Some scientists like big questions—how did the universe begin, what is consciousness, and so forth. But most prefer to take smaller bites, thinking about more modest questions in depth and detail, sometimes admittedly mind-numbing detail to anyone outside their immediate field. In fact, those who choose the larger questions almost always break them down into smaller sized bits, and those who work on narrower questions will tell you how their pursuit could reveal fundamental processes, that is, answers to big questions. The famed astronomer and astrophysicist Carl Sagan, to use a well-known scientist as an example, published hundreds of scientific papers on very particular findings relating to the chemical makeup of the atmosphere of Venus and other planetary objects, while thinking widely and publicly on the question of life's origin (and perhaps less scientifically, but not less critically, on where it was going). Both approaches converge on manageable questions with potentially wide implications.

This strategy, of using smaller questions to ask larger ones, is, if not particular to science, one of its foundations. In scientific parlance this is called using a "model system." As Marvin Minsky, one of the fathers of artificial intelligence, points out, "In science one can learn the most by studying the least." Think how much more we know about viruses and how they work than about elephants and how they work. The brain, for example, is a very complicated piece of biological machinery. Figuring out how it works is understandably one of humankind's great quests. But, unlike a real machine, a man-made, designed machine, we have no schematic. We have to discover, uncover, the inner workings by dissection—we have to take it apart. Not just physically but also functionally. That's a tall order since there are some 80 billion nerve cells that make up the human brain, and they make about 100 trillion connections with each other. Keeping second-to-second track of each cell and all its connections is a task well beyond even the largest and fastest of supercomputers. The solution is to break the whole big gamish up into smaller parts or to find other brains that are smaller and simpler and therefore more manageable. So instead of a human brain, neuroscientists study rat and mouse brains, fly brains because they can do some very fancy genetics on them, or even the nervous system of the nematode worm, which has exactly 302 neurons. Not only is the number of neurons very manageable, the connections between every one of them are known, with the added advantage that every worm is just like every other worm, which is not true of humans—or rats or mice.

"But," says the non-neuroscientist and possessor of a late-model human brain, "my brain and the nematode worm nervous system are simply not the same; you can't pretend to know anything about human brains by knowing about a worm brain, a nematode worm at that." Perhaps not everything. But it is true that a neuron is a neuron is a neuron. At the most fundamental level, the building blocks of nervous systems are not so different. Neurons are special cells that can be electrically active, and this is crucial to brain activity. The ways in which they become electrically active turn out to be the same whether they are in a worm, fly, mouse, or human brain. So if you want to know about electrical behavior in neurons, you might just prefer using one of the 302 identified neurons in a worm versus neuron number 123,456,789 out of 80,000,000,000 in a human brain. The critical step is to choose the model system carefully and appropriately. It won't work to ask questions about visual perception in a nematode worm (they have no eyes), but it is a fabulous organism to ask about the sense of touch (one of the great puzzles of modern neuroscience you may be surprised to learn) because touch is critical for their survival and in the worm you identify the parts that make up a touch sensor by using genetics to literally dissect it. A now almost forgotten statistician and industrialist of the 1920s, George Box, noted that "All models are wrong, but some are useful."

As a quick sidelight this explains modern biology's debt to Darwin. You often hear that contemporary biology could not exist without the explanatory power of Darwin's theory of evolution by natural selection. But it is rarely made clear why this must be the case. Do physicians, for example, really have to believe in evolution to treat sick people? They do, at least implicitly, because the use of model systems to study more complicated ones relies on the relatedness of all biological organisms, us included. It is the process of evolution, the mechanisms of genetic inheritance and occasional mutation, that have conserved the genes responsible for making the proteins that confer electrical activity on neurons, as well as those that make kidneys and livers, and hearts and lungs work the way they do. If that were not the case, then we couldn't study these things in worms, flies, rats, mice, or monkeys and believe that it would have relevance to humans. There would be no drugs, no surgical procedures, no treatments, and no diagnostic tests. All of these have been developed using model systems ranging from cells in culture dishes to rodents to primates. No evolution, no model systems, no progress.

Darwin himself used model systems to frame his questions about evolution—from his famous finches and observations of other isolated island species, to the raising of dogs, horses, and especially the breeding of pigeons, which was popular in his day. Flowers and plants were an especially useful model system because he could cultivate them in his greenhouse. It is notable that Darwin never travelled after he returned from the voyage of the *Beagle*. For a naturalist he was an almost pathological homebody. Many of his insights about origins of species started with "simple" questions about the dynamic and changing nature of these model systems in his backyard—where the light was perhaps better.

There are similar examples of the use of model systems in physics and chemistry and all fields of science. Indeed classical physics, faced with impossible tasks like measuring the weight of the earth, used simplified systems made up of those innocuous balls rolling down inclined planes to measure the stuff of the universe. And post-Einstein physics is even more indebted to model systems, from colliders to computer simulations, for investigating things that happened long ago or far away.

But it is very easy, and very dangerous, to mistake a model system for a trivial pursuit. In the 1970s a US senator named William Proxmire took to presenting what he called the Golden Fleece Award to various scientists whose work was supported by the government, and that he saw as some sort of boondoggle that was swindling the public out of their hard-earned tax money. These Golden Fleece Awards, not limited only to science, but to any government program blatantly wasting taxpayer money, were quite popular in the press and as fodder for satirical comedy routines. Many were well deserved and indeed quite laughable. But in several cases, serious science projects were swept up in the witch hunt. They often had titles that sounded ridiculous when taken literally because they were using model systems. One famous example was the "Aspen Movie Map," a project that filmed the streetscape of Aspen, Colorado, and translated it into a virtual tour of the city. Ridiculed by Proxmire, it later became the basis for Google Earth.

At one time I had a grant from the National Institutes of Health (NIH) to study olfaction, the sense of smell, in salamanders. Aside from wondering why someone might devote his life to this quest, you could imagine many more critical places for NIH dollars to be spent. In fact, I have no abiding interest in how salamanders smell. But I can tell you that the biological nose is the best chemical detector on the face of the planet, and that the same principles by which all animals recognize odors in their environment operates in brains, human ones, to recognize and react to pharmaceutical drugs. Olfaction can tell us about molecular recognition, how we can tell the difference between molecules that are very similar chemicals—the difference, for example, between a toxin and a treatment, a poison and a palliative. And if that's not enough, the neurons in your nose and brain that are involved in this process are unique in their ability to regenerate new neurons throughout your life—the only brain cells that do this. So understanding how they work could tell us how to make replacement brain cells when they are lost to disease or injury. Why salamander? Because they are robust creatures that are easy to keep in the laboratory and they happen to have bigger cells, which are therefore easier to work on, than many other vertebrates. Nonetheless, except for being bigger and being less sensitive to temperature (salamanders are cold blooded), those cells are in the most critical respects just like the olfactory cells in your brain. So am I haunted by a need to know how salamanders smell? No, but they are an excellent model system for working out how brains detect molecules and how new brain cells might be generated. And, by the way, we also get to understand why food tastes good, or not, and how mosquitoes find your juicy body, and how smell plays a role in sex and reproduction.

My grant was titled "Molecular Physiology of the Salamander Olfactory System." Definitely a contender for the Golden Fleece Award, although I think there was too little money involved for me to qualify. But since 1991, when that grant was funded, it has spawned a research program that has produced more than 100 scientific papers and, more important, trained nearly two dozen new scientists. And my case is not exceptional. It is easy to see folly in science: scientists talk funny and can dress weird, and they speak in riddles, literally, because this is what grant proposals are. When you are talking, writing, or thinking about ignorance, it is critical to be as precise as possible. I am interested in understanding olfaction, and chemical recognition and brain cell replacement—but those interests are too broad to be judged on their worth. Of course, they're worthwhile, but how, specifically, would one go about understanding them? It's in the details, and the details often turn out to be funny-sounding titles for grant proposals.

. . .

You may have noticed that I haven't made much use of the word *hypothesis* in this discussion. This might strike you as curious, especially if you know a little about science, because the hypothesis is supposed to be the starting point for all experiments. The development of a hypothesis is typically considered the brainiest thing a scientist does—it is his or her idea about how something works based on past data, perhaps some casual observations, and a lot of thinking typically ending in an insightful and potential new explanation for how something works. The best of these, in fact the only legitimate ones, suggest experiments that could prove them to be true or false—the false part of that equation being the most important. There are many experimental results that could be consistent with a hypothesis yet not prove it true. But it only has to be shown to be false once for it to be abandoned.

So doesn't this sound like a pretty succinct prescription for ignorance? The hypothesis is a statement of what one doesn't know and a strategy for how one is going to find it out. I hate hypotheses. Maybe that's just a prejudice, but I see them as imprisoning, biasing, and discriminatory. Especially in the public sphere of science, they have a way of taking on a life of their own. Scientists get behind one hypothesis or another as if they were sports teams or nationalities—or religions. They have conferences where different laboratories or theorists present evidence supporting their hypothesis and derogating the other guy's idea. Controversy is created and papers get published, especially in the higher profile journals, because they are controversial—not necessarily because they are the best science. Suddenly from nowhere it seems there is a bubble of interest and attention, much like the speculative economic bubbles that develop in commodities, and more scientists are attracted to this "hot" field. There are dozens of examples—is the universe stable or expanding, is learning due to changes in the membrane of the neuron before the synapse or after the synapse ("pre or post," as it's known in the jargon), is there water on Mars (and does it matter), is consciousness real or an illusion, and on and on. Some of these get resolved, while many just fade away after some time in the spotlight, either due to fatigue or because the question gets transformed into a series of smaller more manageable questions that are less glitzy. Newton famously declared, "Hypotheses non fingo (I frame no hypotheses)...whatever is not deduced from the phenomena is to be called a hypothesis, and hypotheses...have no place in experimental philosophy." Just the data, please.

At the personal level, for the individual scientist, I think the hypothesis can be just as useless. No, worse than useless, it is a real danger. First, there is the obvious worry about bias. Imagine you are a scientist running a laboratory, and you have a hypothesis and naturally you become dedicated to it—it is, after all, your very clever idea about how things will turn out. Like any bet, you prefer it to be a winner. Do you now unconsciously favor the data that prove the hypothesis and overlook the data that don't? Do you, ever so subtly, select one data point over another—there is always an excuse to leave an outlying data point out of the analysis (e.g., "Well, that was a bad day, nothing seemed to work," "The instruments probably had to be recalibrated," "Those observations were made by a new student in the lab"). In this way, slowly but surely, the supporting data mount while the opposing data fade away. So much for objectivity.

Worse even than this, you may often miss data that would lead to a better answer, or a better question, because it doesn't fit your idea. Alan Hodgkin, a famous neurophysiologist responsible for describing how the voltage in neurons changes rapidly when they are stimulated (for which he won a Nobel Prize), would go around the laboratory each day visiting with each student or postdoctoral researcher working on one project or another. If you showed him data from yesterday's experiments that were the expected result, he would nod approval and move on. The only way to get his attention was to have an anomalous result that stuck out. Then he would sit down, light his pipe, and go to work with you on what this could mean. But there are not many like Alan Hodgkin.

The alternative to hypothesis-driven research is what I referred to earlier as curiosity-driven research. Although you might have thought that curiosity was a good thing, the term is more commonly used in a derogatory manner, as if simple curiosity was too childish a thing to drive a serious research project. "Just a fishing expedition" is a criticism that is not at all uncommon in grant reviews, and it is usually enough to sink an application. I hope this sounds as ridiculous to you as it does to me. Anyone who thinks we aren't all on a fishing expedition is just kidding himself. The trick is to have some idea about where to fish (e.g., stay out of polluted waters, go where there are lots of other fishermen catching lots of fish—or avoid them since the fish are now all gone from there) and some sense of what's likely to be tasty and what not. I'm not sure you can hope to know much more than that.

It is often said that much in science is serendipitous; crucial discoveries are as much happenstance as the result of a directed search. This makes for nice stories, but it's rarely that simple. As Louis Pasteur, himself a beneficiary of some good fortune, noted, "Chance favors the prepared mind." Lawyers don't make scientific discoveries by accident; only scientists do. That's because their curiosity is driving them to screw around with things to see what will happen. And often, it is true, the thing they find is not what they were looking for, but something unexpected and more interesting. Nonetheless, they have to be looking. The serendipity stories don't teach us that its mostly dumb luck, but rather that we are often not smart enough to predict how things should be, and that it's better to be curious and try to remain open minded and see what happens. Most important, never dismiss anomalous data; it's often the best stuff.

. . .

Now we have a kind of catalog of how scientists use ignorance, consciously or unconsciously, to get a day's work done, to build the edifice we have come to call modern science. It includes a remarkably diverse group of ideas like connectedness, solubility or tractability, and others like measurement, revisiting settled questions, using small questions to get at big ones, curiosity. A jumble of ideas and strategies. Some or all of these come into play at one time or another in the career of every scientist, from one's graduate student days through emeritus (a word my brother pronounces with a long "i," as if it were a disease).

SIX

You and Ignorance

Now we may turn to the question of how *you* can use ignorance to understand that activity broadly called science and the things it produces, rather than being alienated by something you know you depend on. If you meet scientists—at dinner parties, at your kid's school, at alumni events, just by chance here and there while traveling—don't ask them to explain what they do; ask them what they're trying to find out. Scientists love questions. And they usually hate talking about what they do because they are sure they will be boring you out of your eye sockets in no time at all. But they like questions. Ask them what the questions are, what are the interesting things in their field that no one knows about?

For an example of how this might work, we could do what scientists call a thought experiment. Let's say you had the opportunity to spend 5 days with Albert Einstein. What would you do? You could ask him to explain relativity to you. After all, getting it from the master himself would be a unique experience and surely you could come away knowing that you finally got what it is exactly that $e = mc^2$ means, and why it makes bombs work. But you'd be wrong. Chaim Weizmann, the first president of Israel and the namesake of the Weizmann Institute of Science in Tel Aviv, had just this opportunity. He and Einstein were on an Atlantic crossing together, and they determined that for 2 hours each morning they would sit on the ship's deck and Einstein would explain relativity to Weizmann. At the end of the crossing, Weizmann claimed that he was "now convinced that Einstein understood relativity." Weizmann, of course, was no wiser. What he should have asked him was, "What are you thinking about these days, Albert?" "What are the problems you are working on?" "What are the new questions that physicists are asking now that the universe is relativistic, whatever that means?" "What are the loose ends?" And what if Weizmann had asked him questions like those? Then I think Weizmann would have heard an earful of remarkable puzzles and lots of gossip about the new quantum mechanical theories of Bohr and his colleagues and whether this meant that God could be ever be the same God that Weizmann and Einstein grew up believing in. And Weizmann, or at least his thinking, would have been changed forever.

So what makes good questions, and how do you come up with them? And how do you use them to better understand the science? There is a tendency for us to come up with questions for which we think there is an answer, perhaps because ignorance seems embarrassing. But by now you know that this is a bad idea. Ask a softie question of a scientist and you'll just get an answer that's too technical to understand, even if the scientist tries to speak in layman's terms. Francis Crick, Noble laureate and codiscoverer of DNA, admonished scientists to work on what they talk about at lunch, because that was what really interested them. That's often easier said than done for the practical reasons of funding and the like, especially if you don't happen to own one of those Nobel Prizes. But it is the basis of a good question. So ask the scientist you get hold of what he or she was talking about at lunch. That may generate a host of other questions: "What's the one thing you'd like to know about X?" "What is the most critical thing you have so far failed to understand?" "What things (calculations, experiments) aren't working?"

These may seem like general questions that could be asked of anyone or any scientist. But it's not hard to become more specific. You have to do a little background reading to find those questions, but that's easier than you think. You can even start in the popular press—for my class I usually assign a couple of articles from *Discover* magazine or *Scientific American* or even the *New York Times Science* section that are related to the work of the visiting scientist. But even reading science papers, real science papers in real journals, need not be as daunting as it seems. And we often read those as well. There are many scientific papers, even in biology, the field I have a degree in, that are too technical for me to appreciate. But usually I can read the Introductory paragraphs, even in a physics or mathematics paper, and then often I can slog through parts of the Discussion section at the end of the paper. The important thing, I find, is to keep reading past the parts you don't get because of their technical nature. Don't let an unknown word stop you; just breeze on by it. At some point the questions will appear, and you will begin to get what the science is about—the why if not the how.

One of the more remarkable personal experiences I had teaching this class in ignorance was my first try at having a mathematician come talk to us. I was almost as apprehensive as the poor mathematician. Mathematicians are sort of poignant because much of their work possesses an exquisite aesthetic expressing deep truths abstracted to purity, but there are only a few dozen people in the world that they can tell about it.

I read a longish article supplied by the professor on "topology" as it related to the then recent solution of the Poincare conjecture (one of the notorious Hilbert 23). I curled up with the 55-page paper and wondered how much of it I was really going to get through—and how many times I'd drift into stuporous sleep. But it wasn't like that at all. Yes, there was a lot I didn't understand in detail and some of the mathematical notation was beyond me—but a lot of that was easy to get if you just looked it up on the Internet. In the end I really enjoyed, yes enjoyed, reading this paper that opened up a world to me previously unimaginable, where spheres are two- (not three-) dimensional structures, and knots, like those you get in your shoelaces, have unimagined mathematical properties.

The class with the mathematician turned out to be one of the best in 5 years. By the way, that mathematician was John Morgan, then chair of the Math Department at Columbia and now director of the Simons Center for Geometry and Physics at Stony Brook University in New York.

Here are some examples of what have turned out to be good questions in my class:

Do you think things are unknowable in your field? What?

What are the current technological limits in your work? Can you see solutions?

Where are you currently stuck?

How do you talk about what you don't know?

What was the main thrust of your last grant proposal?

What will be the main thrust of your next grant proposal?

Is there something you would like to work on knowing but can't?

Because of technical limitations? Money, manpower?

What was the state of ignorance in your field 10, 15, or 25 years ago, and how has that changed?

Are there data from other labs that don't agree with yours?

How often do you guess?

Are you often surprised? When?

Do things come undone?

What questions are you generating?

What ignorance are you generating?

. . .

Let's review. Science produces ignorance, and ignorance fuels science. We have a quality scale for ignorance. We judge the value of science by the ignorance it defines. Ignorance can be big or small, tractable or challenging. Ignorance can be thought about in detail. Success in science, either doing it or understanding it, depends on developing comfort with ignorance, something akin to Keats' negative capability. Most important, you, that is you the lay person, the reader, can understand an awful lot of science by focusing on the ignorance instead of the facts. Not ignoring the facts, just not focusing on them.

At this point I think it would be helpful to consider ignorance in the particular rather than as a general idea, to get some feel for how it works in the life of a working scientist. To do this, it might be worthwhile to utilize a method of presentation common in medical lectures—the case history—to gain further insight. Can we look at a particular scientist, or a few scientists in a particular field, and analyze their work as a case history in ignorance?

Drawn from my course on ignorance, each narrative that follows is meant to illuminate some particular aspects of ignorance and its importance in the scientific program, but none is a simple parable with a clear and pat message. Like any other life, the scientific life is something of a jumble and the process for each person is, as I have said, idiosyncratic. I have tried to emphasize the points of ignorance in these narratives, but I have not purposefully slanted them to be examples of this point or that. I think you have read enough about scientific ignorance to appreciate it in its various guises wherever it pops up, as it does aplenty in these little histories.

SEVEN

Case Histories

1. IS ANYONE IN THERE?

Is there anything harder to know than what's inside another person's head? What is he or she thinking, feeling, perceiving? Is my "red" her "red"? What is it like to be him? Is there anything we can know with less surety?

Yes. What is going on inside another *animal's* head.

And this is where Diana Reiss looks for ignorance.

Cognitive psychologist Dr. Diana Reiss wonders whether other big-brained animals have higher mental faculties similar to ours. Dr. Irene Pepperberg at MIT is asking the same sort of question of an animal with a much smaller brain—a bird brain, in fact. The deeper question that both of them are asking is whether there is a smooth progression of mental function across species, or is there a mysterious discontinuity when it comes to humans? Can we see into the animal mind? Is there a mind there to see? For many years, many centuries really, it has been dogma that animals and humans are fundamentally different when it comes to cognition, to mind. It may be that our hearts, livers, kidneys, and other parts are all recognizably similar, if not precisely the same; it may be that our physiology and biochemistry are fundamentally the same; it may be that our reproductive and eating requirements are pretty much indistinguishable. But when it comes to mind, there is a difference. Historically this difference was inextricably bound up with the notion of the soul—something humans clearly have and other animals probably (we hope?) do not. For many people this is still the crux of the matter. For a scientist this is no longer where the questions endure.

As part of the inexorable march toward proving the "Law of We Ain't Nothing Special," it now appears that in addition to not being the center of anything cosmological (solar system, galaxy, universe, multiverse …), we are also not so special among the living creatures inhabiting our little, out of the way, dirt sphere. Our brains, although bigger than most (but not all) and of perhaps more complex organization (except that we haven't really looked as carefully at other complex brains), are still fundamentally more similar to, rather than different from, those possessed by at least the rest of the order of mammalia.

The trouble starts when you talk about consciousness—or where conscious awareness resides. In the 4th century B.C. Aristotle worried about this difference, arriving at what he called (and we have via Latin translation) the "Scala de Naturalia" or Ladder of Nature in living things. In this scheme, plants have a vegetative soul for reproduction and growth, nonhuman animals possess, in addition, a sensitive soul, which through the senses perceives the world, and humans add to both of those a rational soul, for thought and reflection.

Whether soul or consciousness, it is no longer in the breast of man, no longer in the heart, no longer even in the pineal gland (as Rene Descartes proposed)—it has come to rest in the brain. It is the "special sauce" that makes the human brain conscious, more than just a fancy computer that runs the organism in predictable if fascinating ways. If animals have brains like ours, do they also have souls? Even if their brains are only half as good as ours, don't they get some soul for that? Do they have feelings? Can they be hurt in more than just purely physical ways? Do they *feel* pain, not just experience it? The way you think about these questions will affect your moral view of the world on everything from getting vaccinated, to eating meat, to aborting pregnancies, to worrying about the climate, to thinking about death.

Except that we don't burn folks at the stake anymore, the stakes can be high when these questions arise.

Descartes, near the very beginning of what we identify as the Western scientific tradition, stopped the field of animal cognition in its tracks by claiming that most animal, and even much of human, behavior is like the workings of a machine, predictably adhering to the laws of mechanics. Thinking was something separate from the rest of behaving, maybe even governed by different laws and principles. While this extreme view, known as the mind-body duality, is no longer commonly held, precisely where we draw the line, or if there is a line, between mind and behavior remains very controversial.

Here we have science fraught with history and carrying the baggage of possible religious, or at least moral, connotations. Reiss points out that the threshold for showing cognitive abilities in animals is much higher than it is in humans, even obviously damaged humans with severe mental dysfunction. No matter how retarded a child may be, we still believe he or she has essential human qualities, including a cognitive life that is soul-like. Animals, on the other hand, have to perform at nearly superhuman levels to be even considered as having something we might call "mind," whatever that is.

In fact, this is precisely one of the big problems for Reiss. What we call mind tends to be circularly defined

as something that humans have. But this kind of definition, even if only implicit, is useless. It creates ignorance
in precisely the wrong way—by appearing to mean something, when in fact it means nothing. This has the effect
of stalling inquiry rather than propelling it. As Reiss asks,
"Why do we think animals don't think? We begin with
a negative starting assumption and then must prove that
they do."

Even worse perhaps is that there is an implicit double
standard in the thresholds for what is considered proof
and how the data are to be obtained. This is what the late
Donald Griffin, a Harvard researcher in animal behavior
who discovered the sonar abilities of bats, called "paralytic
perfectionism"—setting the standards so high that progress
is virtually impossible. For example, in the several projects devoted to teaching chimpanzees rudimentary language skills, much of the criticism has been over charges
of "cueing"—the conscious or even unconscious cues that a
researcher may give to a subject that changes its behavior—
what many of us call body language, although it may not be
limited to posture. Now, of course, these sorts of social cues
are critical to how a child learns language. Imagine trying
to teach your son or daughter English without ever smiling
or nodding or changing your expression—just rewarding
him with a cookie when he says something correctly and
ignoring him when he makes mistakes. This borders on

child abuse, but it is required procedure in animal language experiments.

But as Reiss points out, this apparent injustice serves some purpose as well, helping to avoid the many pitfalls that come along with cognitive research and collecting data about another mind that finally you can never know entirely. For example, the outward appearance of cognitive behavior may not be an accurate depiction of inner life— even in humans. We have expectations about the behavior of others, built into us either by genes or early learning, and these expectations are more likely to be satisfied by what we observe. We have an idea of what consciousness looks like, and we are apt to recognize things that look that way and call them conscious behavior—even when they are not.

The famous story of Clever Hans the horse is instructive. Clever Hans was a horse that could apparently perform mathematical calculations. Owned by a retired schoolteacher, he received wide attention from the press and the general population—possibly because of an interest in animal intelligence motivated by the recent publication of Darwin's *On the Origin of Species*. Herr Van Osten, the school teacher/owner, would ask his audience for a mathematical problem—how much is 5 + 3, for example—and then ask Hans for the answer. Hans, to the amazement of all, would tap his hoof eight times. He was just as good at addition, subtraction, multiplication, division, and other simple numerical tasks. Hans was a sensation. Investigations by panels of "experts" concluded there was no fraud. He performed scores of free demonstrations to crowds all over Europe.

Finally a young graduate student in psychology, Oscar Pfungst, set up a series of experiments that revealed the method in the cleverness—and it was surprisingly subtle. Pfungst found that Hans could perform well no matter who his human interlocutor was—van Osten, strangers, or even Pfungst himself—as long as that person knew the answer. If the questioner was in the dark, so was Hans. Hans would also fail if he couldn't see the person—for example, if they were separated by a partition or if Hans was wearing blinders. This led Pfungst to the realization that Hans had to be getting some cue from the person, and by careful observation—of the person, not the horse—he found that people would tense the muscles of their body and face at the beginning of Hans's answer and release the tension when he arrived at the correct hoof tap. Hans was clever alright, just not at mathematics. Hans was reading very subtle changes in posture, expression, and attitude in his human collaborators. Most remarkable, Pfungst discovered that even after he knew he was providing these cues, he could not consciously prevent himself from doing so. If he knew the answer, he would involuntarily alter his demeanor in ways that the very clever Hans could observe.

This realization changed the course of experimental psychology, and, for that matter, any field having to do with living organisms, forever. In large-scale drug testing, patients get either the real drug or a fake one, the placebo, and the administering doctor cannot know which one the patient is getting. Otherwise he or she would unconsciously cue the patient. Even the person supplying the drug to the doctor cannot know, because the doctor will figure it out and communicate that to the patient. And all of this can, and mostly does, happen unconsciously.

The method that scientists use to control for these Clever Hans effects, as they have come to be known, is the "double blind." That is, either the experimenter cannot know the correct answer or she cannot be available to the subject. The experimenter herself cannot be trusted to hide the correct answer because she will give it away involuntarily. This is all well and good for drug trials, but it creates a serious problem for the cognitive researcher. The very social cues that are believed to be critical to the complex task of communication, especially linguistic communication, are those that need to be ruthlessly removed from the experiment by a double-blind procedure. But removing the social aspects of the process destroys the experiment. This is the double bind of the double blind.

It is true that the various language experiments with chimpanzees, and the few with dolphins, were ultimately failures in the sense that researchers never taught any of them to sit down and describe what life as a chimp or a dolphin is like. But they were groundbreaking nonetheless because they revived the subject of animal cognition as a scientific question. They led to the recognition of tool use, symbolic behavior, numerosity, empathy, and even self-awareness in animals previously thought of as mere behaving machines. Beginning in the late 1980s it once again became okay to investigate what was behind the behavior of animals, and to consider that it was more than just gears and levers.

But how precisely do you ask questions about mental activity in an animal? Reiss and Pepperberg take two different approaches in the particulars, but they are fundamentally the same strategy, based on the same guiding principle. That principle is one that is difficult for many scientists to swallow, because it relaxes control, gets the experimenter out of the driver's seat, and leaves it up to the subjects—dolphins, elephants, and parrots in this case—to produce the results. For both Reiss and Pepperberg, the key strategic leap was to give up worrying about the definition of consciousness or self-awareness, about what the thing (if there is a thing) was and how to produce it, and instead to provide an opportunity for an individual creature to simply show us whether it acted consciously. This may seem a subtle distinction, but it is what has allowed Reiss and Pepperberg to advance. This is an example of the question, the right question, asked in the right way, rather than the accumulation of more data, that allows a field to progress. As Reiss says, "Our only chance is to get these occasional glimpses of a mind at work." It will not reveal itself in some cagey test. These have been tried before, and in each case some Cartesian-minded behavioral psychologist has been able to show that what looks like conscious, self-driven behavior can be just as easily replicated with simple stimulus-response schemes. No need to invoke consciousness to explain some apparently complex behavior, simple reward systems can do just as well. And you won't see consciousness in a magnetic resonance image (MRI), because you can't find it in human MRIs. There is no seat of consciousness, no bump on the head, and no deeper brain structure. It's an "emergent phenomenon" that appears in some creatures and not in others. Who has it and why?

I really like this idea of a "glimpse." Like the notion of dis-covering, but even more humble; it is often all that is available. Engineering them, glimpses, is the subtlest kind of experiment that one can design. Facts don't often stand still, and they are often only perceptible to peripheral vision. It's hard to know (i.e., predict) where they will come from and when.

For Reiss, the key is watching patiently and giving animals opportunities to demonstrate their abilities. She waits for the occasional "glimpse" of another mind at work, hoping that it will be recognizable. This notion of stealing a "glimpse" into the question is critical to Reiss's approach. It is almost Zen-like. Looking directly at the thing causes it to disappear, and being too active creates only what you want to happen. She works to create instead an opportunity while remaining dispassionately focused. Do you see the paradox that is the life of this scientist?

One practical problem with this strategy is that the numbers can be small. Indeed, often the data are anecdotal, they are a story, but with a little luck, one that perhaps helps to design an experiment. Reiss has a classic example with a dolphin subject that was part of her graduate work.

As a first step to establishing a relationship with her subject animal, a female dolphin named Circe in an aquarium in southern France, Reiss took over feeding her. Feeding allowed her to establish some training parameters that would come in handy later. The most basic of these is something called holding station—it means simply to come stay here with me and focus on me. Reiss would hold her hand out, and Circe would come from wherever in the tank she was swimming and poke her head out of the water, for which she would get a fish. More precisely, she would a get a piece of fish because she was used to having her fish cut up into three sections—heads, bodies, and tails. Circe didn't care for tails and would reject them—effectively training

Reiss to only feed her heads and bodies. Now when Circe refused to hold station or to do some other task that was being requested of her, Reiss would not only withhold the food reward she would back off about 10 feet away from the tank for 5 minutes in what is known as a "time-out." This time-out is not much different from what teachers use on misbehaving young children—it's a kind of punishment because it means the offender, student or dolphin, cannot do anything to make it right or to get the reward. It's a very effective strategy that has the added advantage of being physically painless.

One day Reiss was conducting an experimental session and Circe was working away at performing some test, when Reiss inadvertently fed her a tail—the nonpreferred part of the fish—that had slipped into the bucket. Circe spit out the tail, promptly turned away, and swam back 10 feet from the station, turned around, and, the top half of her body out of the water, watched Reiss standing there bucket in hand but nothing to do. She had given Reiss a "time-out."

Or had she? For Reiss it was unmistakable. But it was once. It was an anecdote. Can an experiment be designed from this? Can an anecdote be turned into an experiment? As Reiss says, "While we are trying to learn about them, they are trying to learn about us, and that may be the most interesting part of the experiment." The situation is fluid; there are too few controllable variables. And yet every so often there is a glimpse that reveals a mind at work—unmistakable but also unquantifiable. It is these glimpses that Reiss works for. (In fact, Reiss staged a few more "mistaken" feedings of tails and each time Circe punished her with the same behavioral response—a time-out).

So the problem is how to create a moment when you get a glimpse. This is a not uncommon theme in science. Discoveries don't come every day. Even after you settle on the question, there is still the business about how you will pursue it.

Currently, Reiss is using mirrors to get a glimpse of what animals think about themselves—if they think about themselves. By the age of 18 months all humans get the mirror illusion (that's what it is after all), and since the pioneering work of Gordon Gallup Jr. in the late 1970s we know that chimpanzees and other great apes also figure it out. This is one of those experiments that seems painfully obvious—in retrospect. Once done, nobody can understand why it hadn't been done decades earlier. Gallup, wondering whether chimpanzees, given the opportunity, might show self-recognition, simply provided some chimps with mirrors and observed their behavior. There was a clear evolution of behaviors from social (as if the image in the mirror were another chimp) to contingent (doing something and seeing whether the mirror guy imitates you—the classic Lucille Ball, Harpo Marx comedy routine) to self-directed (for example, using the mirror to inspect the inside of your mouth). So with the right opportunity the chimpanzee showed us that it can figure out the mirror. But the ultimate test is whether the chimpanzee "sees" this image as itself, in the sense of understanding it is itself. For this, Gallup devised what has famously come to be known as the mark test. Lightly anesthetized chimps were marked on their foreheads with a red dye and upon awakening were provided access to a mirror. When they got around to looking in the mirror, they noticed the red mark on their foreheads, touched it, and investigated it using the mirror. Indeed, the image in the mirror was herself. They are our cognitive cousins.

But monkeys don't get it, nor do dogs or cats or other very smart species. Is it just higher primates? Is there something specific about primate brains? Are we special after all? Do we just have to admit chimps and gorillas into the club to keep us special? Reiss had the mind of another smart species in mind; she wondered what dolphins might do with a mirror. Why? They have big brains, about the size of ours in relation to body size, but they are in every other way completely foreign. They live in the water and move easily in three dimensions, they have no hands, they have no facial expression except that frozen Mona Lisa smile, and in a host of other small and big ways they differ from the typical land mammal and in particular from primates. They are, as Reiss jokes, extraterrestrials (at least in the narrowest sense). Their last common ancestor with primates lived 60 million years ago. If they can do the mirror thing, then it is clearly not a primate-specific mental trick.

The results of Reiss's work are published and you can read the paper (Proceedings of the National Academy of Sciences, 2001), but the short answer is that dolphins do recognize themselves in a mirror, going through all the classic behaviors and phases of recognition seen in humans and chimps, including finally the mark test. And while the results in that paper are very striking, the real value in this work is the many questions it has generated because it has expanded the question set from what makes primates special to what the requirements are for a brain to develop a mind. What deeper commonalities are there between self-aware species? Why has self-awareness arisen in such different species? The demonstration of mirror self-recognition in dolphins generates more questions than it answers. In fact, all it does is generate questions. What a great experiment.

But Reiss wasn't finished yet. Along with Frans de Waal, her colleague and noted primatologist, she collaborated on a mirror self-recognition test of elephants. Elephants weren't chosen out of the blue, so to speak, but because they showed some behavioral characteristics that de Waal and Reiss thought might predict that they would pass the mirror test. Once the exclusivity of the mirror self-recognition club had been broken by dolphins, it was reasonable to consider other characteristics than being a primate that would indicate strong sense of self. And, indeed, these large, and large-brained, animals needed little time to go through the now predictable behaviors leading to correct mirror use—social, contingent, self-directed—once a mirror, a very large, very strong mirror, was placed in their enclosure at the Bronx Zoo in New York City. Like the dolphins, the elephants were rapidly attracted to the mirror—perhaps because it was a new thing in an otherwise not so interesting environment that they were all too familiar with. Whatever the initial attraction, the appearance of oneself in the mirror was quickly recognized and tested by these large-brained mammals.

Reiss thinks this testing may be part of the clue to determining what makes an animal mirror-self-conscious. Remember, these experiments were designed as an opportunity for animals to use their minds in ways we could interpret. What Reiss sees is that animals that come to recognize it's them in the mirror are already highly inquisitive. They don't just take note of things in their environment; all animals do that. And they don't just learn cause and effect the way Skinner's pigeons do; all animals learn from experience. These species, the mirror-savvy ones, "test contingencies," as she likes to say. "They actively probe their environment looking for effects from causes they instigated. They are scientists." Many animals check out the mirror when they first see one but rapidly dismiss it as having no effect. Some animals probe more deeply. Those are the ones that interest Reiss. Those are the animal scientists.

Now the mirror-self-recognition fraternity in Reiss's experiments included at least two nonprimate species that also had little in common with each other. So what was the common denominator? Is there one? Are there many? Has self-awareness evolved many times independently, like color vision? Is it perhaps not as rare as we once expected? Does it really indicate self-awareness, in the same way that we feel self-aware? That is, does it extend beyond seeing a version of yourself and simply taking advantage of the new information? Why would mirror-self-recognition evolve in the first place, even in humans?

For Reiss the mirror is a kind of glimpse machine; it provokes the mind of the animal and provides a glimpse of what may be going in there. Can we locate the part of the brain that is active during mirror use? Is it just one place, or is it distributed? Will it be something that only certain brains possess? What happens when you distort the image in the mirror—can other animals "get" the funhouse mirror metaphor? Will they still recognize themselves in a distorted view? How far into the mind of another can a mirror take us?

Reiss wants to try the mirror test on an octopus, a very visual, surprisingly intelligent *invertebrate*. What a

Pandora's box of questions that could raise. "Yes, indeed," Reiss smiles.

Dr. Irene Pepperberg used a technically different method to work with an African Grey Parrot named Alex, but with precisely the same purpose. She taught Alex a vocabulary of some 100 words to describe a variety of objects (blocks, fabric, food, etc.) and qualities (color, texture, number, etc.) and then allowed Alex to make use of these linguistic tools to manipulate his environment. Like Reiss, Pepperberg wanted to give another brain a chance to show what it was made of, in this case a bird brain. She has recently published a very accessible book on her work with Alex, so I won't go into detail here. Rather I want to make the single point of how her strategy courts ignorance. The purpose of the training was not to show that a bird could produce utterances that appeared to be linguistic, because then we would just have to have a lot of maddeningly circular arguments with Noam Chomsky about what constitutes "true" language. Rather, all the training was to give Alex a chance to let us glimpse his brain. Finally, the purpose was to be able to frame better questions about higher brain function. Alex was a model system for consciousness. No, he wasn't as complex as a human, but he turned out to have more abstract mental powers than you might have expected. Alex learned to count (really, not like Clever Hans) showing a sense of numerosity that he then applied to getting more of his favorite things; Alex coined new words by combining existing phrases he knew to express new thoughts. Before Alex, who would have thought you could have a model system for consciousness? By breaking down our prejudice that consciousness is something made of whole cloth that only shows up in humans, Pepperberg's work enables us to ask more detailed questions about what consciousness is and what it is made of and when and why it shows up. A talking parrot is not news; one who thinks about what it says is. Of course, that might be true for humans, too.

Alex died in April 2010, suddenly, at the age of 35, a devastating blow to Pepperberg's research. This is one of the great difficulties in research of this nature, where the subject is the limiting factor. In biochemistry, for example, it is the death of the researcher that interrupts the research—but they are easy to replace. Easier than a thinking parrot for sure.

Irene carries on with new birds, but the amount of time and the intellectual and emotional effort dedicated to Alex will never be recovered.

The take-home message from this case history is not just that scientists design an experimental strategy based on what they don't know, but that the truly successful strategy is one that provides them even a glimpse of what's on the other side of their ignorance and an opportunity to see if they can't get the question to be bigger. That's progress.

2. EVERYTHING = 1(TUTTO = UNO)

Why does there have to be a unified theory of everything before physicists will be content? There is no comparable drive in chemistry or biology. Are those fields fundamentally different or perhaps not as mature as physics? Physics is probably the most remarkable success story in science in its ability to explain everything—or almost everything. And perhaps there's the rub: there seem to be just a few bits missing, and that's sometimes worse than having whole swaths of ignorance. Physicists know a lot about how things work when they are very small and have almost no mass— this is the quantum world; and they know about the fundamental characteristics of the very big, the cosmologically big—this is relativistic physics. But they don't know how to connect these two universes—this is the elusive unification. Of course, it may also be that the grand unification, if achieved, will immediately reveal more parts that are suddenly inexplicable—much like establishing the atom, the "indivisible" unit of mass, almost immediately gave rise to the recognition that there were constituent parts previously unimagined. Thus, even now there are the nefarious-sounding "dark matter" and "dark energy," unseen (in the deepest sense of being undetectable directly) but making up the bulk of the universe, and that may or may not be part of the grand unification.

So what happens if these two physics can't be fused? Well, then there are numerous key questions that physicists will not be able to address, some of which they know already and perhaps some others of which they have only a hazy inkling. Many of these questions have to do with some fairly fundamental issues—the true nature of mass, of time, of the beginning of the universe, of why it is like it is. At this juncture, physics intersects cosmology in a field that has come to be called astrophysics, where asking questions about the universe out there, way out there, turns out to be useful for asking questions about physics right here. The astronomical universe has the kind of laboratory conditions that could never be found on the earth. To the extent that we can observe what's going on out there—and to the extent that we are confident that physics in the region of the earth and this solar system is no different (that is, obeys the same laws) from what it is at the farthest boundaries of the universe—then the physicist can pose fundamental questions about the nature of matter and energy in the laboratory of the cosmos.

Sounds great, but here are some serious problems associated with being an astrophysicist who uses the universe as laboratory. One is that you can't actually get there. Two is that you can't do experiments on "it." Three is that you are a part of it, so measuring it objectively can be awkward. Four is the time limit, the speed of light, which puts a horizon on time so that you can't see there, or then either. Actually, you can only see "there" "then"; that is, looking out into space is looking back in time, so the whole history of the universe is laid out in front of you—look out the right distance and you can see whatever "then" you want (just not "here"). This maddening confusion of time and distance is only one of the mind benders that cosmology offers. Five is that there is only one of it, so you can't have a large sample size. And just in case any of those get solved, there are more problems waiting.

Well then that seems to be more than enough ignorance to go around, which should by now suggest to you that this is an ideal system. In fact, astronomy and astrophysics have provided tests, proofs, and solutions to problems ranging from gravity to calendars in the ancient world to the chemical periodic table to Einstein's general relativity. Their power is undeniable. From the unseen ignorance they spew out proofs and data like a dark neutron star throwing off particles. Astronomy is a true laboratory because it is home to deeper and deeper layers of questions.

Scientists who relish these unanswerable questions divide up into two main approaches: theoretical and experimental. This division, once very competitive, is no longer so distinct, as they have become mutually reliant on each other. It is still true that theorists mostly manipulate mathematical equations and experimentalists collect data, but theorists depend on experimental data to support assumptions, validate predictions, and in some cases distinguish which of several mathematically equivalent models is physically realistic. And experimentalists depend on theorist-derived models to design experiments and to interpret data. For this case history we have three scientists who span the range from theoretical to experimental, but I caution the reader not to be too parochial about these definitions.

Brian Greene is a theoretical physicist who is also well known for his elegantly written descriptions of the hard-to-get-your-head-around concepts that make up the Alice in Wonderland worlds of relativity and quantum mechanics. He is a unifier. His physics, which is primarily expressed in mathematics, is a search for the true description of the universe we inhabit based on what we observe and some guesses about what we should be able to observe if we had the technology. And this universe is likely to be quite unintuitive. He believes it won't be complicated, in the sense of having lots of moving parts to keep track of; it will be sophisticated and it may use difficult mathematics, but it needn't be complicated.

One of Greene's strategies is to try to work out the metaphorical explanation of the mathematical insights that are at first so other worldly. "To get over experience," as he puts it, is to develop an intuitive (that is to say, unintuitive) view of relativity and quantum mechanics. This may not be as improbable as it sounds. You could imagine that if Einstein had come before Newton, we all might have a different view of the world. It would be normal to think of time as malleable and gravity as a geometrical feature of space. It might then seem that our everyday Newtonian views of gravity, with its massive bodies "attracting" each other, are hopelessly complicated (as many high school physics students will tell you) and full of arbitrary-seeming formulas. Greene works hard at this, at understanding the world from uncommon and unintuitive points of view. You might say this ability was the key to Einstein's breakthrough: he was willing to get past experience, in this case a Newtonian perspective, and imagine life as a photon riding a beam of light.

How counterintuitive might Greene's universe be? He thinks that the ideas of space and time will not appear in the fundamental description of the universe. These will be derivative concepts in the way that temperature is a derivative unit—it is a result of molecules in a gas or liquid moving fast, but one fast-moving molecule doesn't have any heat. Thus, even though his favorite theory is based on strings occupying 11 dimensions, they may not include the familiar 4 of space and time. We don't even have the language to ask the questions without invoking paradoxes and logical inconsistencies. "Time will disappear at the beginning of the universe," but what then do we mean by "beginning"? In an improbable conflation of Bill Clinton and Gertrude

Stein, you might ask of the beginning of the universe, "Is 'is' there?"

How does one think thoughts that are so far outside our language that framing them is difficult and expressing them to others invites frustration for both speaker and listener? Well, there's mathematics, because it doesn't have the same conceptual limits of spoken languages. In mathematics there are computational rules and you follow them and get the results that you get. Sometimes you are forced to make some pretty strange assumptions in the math, because if you don't, you get results that are not interpretable—they're mathematically correct, but meaningless. So Greene and his colleagues do math, make assumptions that sometimes seem fanciful but that they nonetheless have to defend, get improbable results, and then try to understand how these could describe the universe. All in a day's work. Although most days it doesn't work, and they go back and attack from another angle, with another set of mathematical strategies. They get their glimpses through equations that challenge and force their imaginations.

In fact, Greene has a somewhat cavalier approach to all of what he does. This is especially striking because he is the very public face of the controversial String Theory model, and you might think he has a lot invested in its eventual success as the prevailing model for unifying physics. And while he is quite confident that it will work out, that he will be proven correct, that he is on the right track—he is ready to be shown otherwise, to be completely wrong, to find that he has been chasing down one of those cats that aren't there. "If string theory is wrong, I want to know right away"; no sense wasting any more time on it. Of course, one is almost never completely wrong or irrelevant. Just as one is rarely completely right. Engagement with the universe is the critical issue—right and wrong are for gamblers, politicians, judges, and the like. Science lies elsewhere, with the engagement. From the engagement comes the unexpected, the sudden intuition, the abruptly obvious comprehension, and the new ignorance.

Astronomer David Helfand, more on the experimental than theoretical side, is waiting for a nearby star to blow up. When I say nearby, it is important to realize that terms like "near" and "far" in astronomy are not the same as when you give directions to your friend's house for a party. Helfand is hoping for something local, here in the Milky Way, in our hundred thousand light year neighborhood. An especially bright supernova appeared in 1181, and it was recorded as a "guest star" in religious and historical reports and personal diaries from China and Japan. (As an interesting sidelight on what we don't know we don't know, the event went largely unnoticed, or at least unrecorded, in Medieval Europe, where Aristotelian views of a never-changing perfect celestial sphere still held sway. A fairly bright new star in the sky was apparently not considered an important enough event to record; it wasn't likely to be any more than some "disturbance." You see how easy it is to miss things. In contrast, another supernova that could be detected by the naked eye in the night sky occurred in 1572. Now well into the Renaissance and more liberated from the tyranny of classical authority, it was recorded by astronomers throughout Europe. Indeed, this one was used as a case against the Aristotelian view of immutable heavens.) Stars blowing up leave traces that can tell you something about what they were made of and how they operated and why they came to an end. A stellar death is a kind of astronomical crime scene, and the science uses a forensic approach, analyzing what's left behind to reconstruct and understand the causes of an event that couldn't easily have been predicted beforehand. The critical thing is to know exactly when the star exploded, the time of death if you will, because interpreting all the rest of the evidence will depend in one way or another on that. And it also helps to be there as close to the event as possible. Right now Helfand uses historical reports of stellar mishaps that appear in ancient texts to date his explosions, but to be there when it really happened, to have the telescopes and instruments trained on the star within seconds of it actually blowing up, that would be the real prize. (There was a supernova in 1987, but, unlike the 1572 event, it was in another galaxy and too far away to be quite as useful as one here in our neighborhood.) The data that could come from that sort of observation would clear up many nagging questions. How fast do things cool off, that is, dissipate energy? What happens to atomic nuclei and their quarky constituents under conditions that could never be replicated on earth? What happens to space and time in the immediate region of an exploding star? All great questions. It's also clear, because he can barely suppress it, that Helfand would simply like to have the astronomically good luck to be around when one of those epochal events takes place. Who wouldn't?

Of course, one probably has, but we don't know about it. For example, the red star that makes up Orion's right shoulder (Helfand admits it's the only constellation he knows) is ready to go, but since it's 400 light years away we are seeing the star that was there when Shakespeare was producing plays. It may have actually blown up the night Hamlet was premiered, but word (or rather light) of it will not have reached us just yet. The strange thing about the universe is that we know better how it looked in the past than how it looks now. Betelgeuse may be long gone in a spectacular fireball, but we are none the wiser. In the deep universe we only see the past; we can never see now.

As it is, Helfand, like many astronomers and cosmologists, lives in a warped time zone where things happen very slowly and then very rapidly and often without much warning, even though they are sometimes cataclysmic. For example, Helfand studies an astral event that is just over 800 years old, using techniques that confine him to making observations every 10 years or so (depending usually on the launch of some new satellite with new instruments), and then the data stream out in millions of bits per second and can sometimes be interpreted in minutes, followed by long days to complete the deeper analysis, and weeks or months to publish it.

Something not often appreciated is the tremendous range of the time scale of science. Scientists study some events that unfold over eons and some that last millionths of a millionth of a second. Although science continues over generations, it is rare that an individual scientist picks a problem that cannot be resolved within his or her lifetime. Helfand admits, sardonically but perhaps a little sadly, that for one of the current problems he is working on he now realizes that he will likely die without the answer, simply because NASA changed plans and now doesn't expect to launch the required instrumentation for about 30 years.

How much does our 50-year working human life span govern the way scientists organize their questions? I think it must, but most of us rarely consider it consciously when we talk about experimental programs or when we identify the important questions that we would like to work on. Those longer range goals get a brief mention at the end of review articles, they are the futuristic imaginings, but they have little effect on the day-to-day work of most scientists.

Helfand makes a quick list of the questions that astronomy could impact in the next 400 years, clearly not concerned with his life span or anyone else's. These are the centers of ignorance: nuclear matter at densities greater than those in a normal nucleus; the identity of dark matter; the polarization of the microwave background radiation leading to measurements of quantum fluctuations (more on this scary-sounding phrase later); the reconciliation of general relativity and quantum mechanics (that's the big unified field theory); the identity of the dark energy; gravity waves as a possible test of string theory and higher dimension universes. Quite a list. I started this section by saying that physics could be considered among the most successful of the sciences for all that it has explained. It could also be the most successful for all the questions it has spawned. That list of Helfand's couldn't have even been conceived by some of the great minds in physics a mere hundred years ago.

Amber Miller, a true experimental astrophysicist, practices a kind of archaeology of the universe. She looks for fossil traces of the very early universe in the remnants of the initial explosive event that brought the whole thing into existence. Using these fossils, she asks detailed questions about the early universe so that we can understand why it has come to be what it is, whatever that is.

The numbers that astronomers use are big—everything is unimaginably vast out there. It's one thing to hear these numbers, and see them plugged into the theorist's equations, often using special notations that allow you to state a big number without actually writing out all those zeros. This also makes it easier to overlook their true size. The number 10^{21} simply doesn't register on the mind in the same way as a 1 followed by 21 zeros: 1,000,000,000,000,000,000,000. But for an experimental astrophysicist, someone who actually makes the measurements and builds equipment to specifications that are determined by those numbers, it all becomes more real—or curiously less real because you realize how truly incomprehensible these distances and times are.

Miller is desperate to find out what happened at the very beginning of the universe. The very, very beginning—say within the first 10^{-35} seconds of its existence. Now that's a very small number, but in the mirror world of astrophysics that means looking out at a great distance, a very great distance, because in astrophysics distance is time. Light from those very early moments in the universe must travel to us from what is now the very edge of expanding space. Long ago means far away.

The universe creates space ahead of it. The first explosion, the Big Bang as it is now called, created time and space and as the edge of that explosion moves outward it continues to create time and space. The universe is not so much expanding; rather space is being created and the universe, the things in it, galaxies and the like, are simply filling up the expanded space. Talking about these things reminds you how important it is to ask the questions properly. No sense asking what's on the other side of that expanding edge of the universe. It isn't there; it hasn't been created yet. Just because a question can be asked doesn't make it a meaningful question.

In trying to understand what is happening on that edge of creation, Miller looks for cosmological fossils by sending balloons into the outer atmosphere loaded with sophisticated instruments that can measure something called the cosmic background microwave radiation. In the 1940s it was proposed that if there were a Big Bang–type explosion that started the universe, then even all these billions of years later there should still be some trace of it permeating the universe. This trace was calculated to be a very low-frequency hum, like noise on your radio when you can't quite get the station. In one of the great stories of serendipity in science, precisely this noise was discovered by two scientists at Bell Labs while they were trying to get rid of a troubling and persistent hum that plagued a new radio-telescope instrument they were testing. That hum was not a fault in the instrument but the cosmic background radiation left behind by the Big Bang:—a 13.7-billion-year-old fossil.

What this fossil shows is a bit curious and leads to a conundrum in cosmology. The universal hum, the cosmic background radiation (CMB), is nearly the same in all directions. This suggests that things out there are pretty homogenous, which would be okay but remember when we say "out there" we are really saying "back then" because in astrophysics we can only see what has had time to reach us traveling at the speed of light. This is why astronomers measure distance with a unit called a light *year*. In this Red Queen world it is easier to see the past than the present. Looking at "back then" leads to a subtle question. The universe we are seeing now is much larger than the universe we could see some time ago. Even in the last 10 years the observable universe has increased in size by a radius of 10 light years, so the parts of the universe we are seeing now couldn't have been seen 10 billion years ago. If they couldn't have been seen then, they couldn't have interacted with each other. So how come they're the same now? It's as though we have a condition now that must have been caused after the parts were able to interact—that is, the result seems to have come before the cause. Not sensible at all. Smacks of the Red Queen telling Alice that she could believe six impossible things before breakfast.

The only current solution to what is called the "horizon problem" was first proposed by Alan Guth and colleagues of Stanford, in 1980, who suggested that shortly after the initial explosion there was something called the inflationary period, where the universe suddenly entered an accelerated expansion, and then once again slowed down into the more or less constant expansion that we observe today. In that case things that were close together were suddenly pulled apart in this faster-than-light expansion and now find themselves even farther apart as the universe has continued expanding at a more "normal" rate. Because this happened so soon after the initial event (somewhere in the first 10^{-35} seconds), it may be that the entire observable universe we find ourselves in now was no more than a 1 centimeter patch at the moment of inflation. I should say that this is not just some idea pulled out of a hat, although it may sound that way. There are good reasons to believe that in the very, very early universe conditions were ripe for this sort of accelerated expansion. An inflationary period in the evolution of the universe solves numerous problems besides the "event horizon" and is now generally accepted in one form or another, but the proof is admittedly still thin.

One of the main proofs for the inflationary model has come from the work of experimental physicists like Amber Miller, measuring with exquisite precision the cosmic background radiation that presumably arose just after this accelerated expansion when the universe had its "hot big bang" and separated matter from energy. What Miller and others have found may at first sound like the opposite of proof because what they have detected are very tiny disturbances in that smooth background radiation. These are called anisotropies, an unfortunately difficult-sounding word standing for a relatively simple idea that means that the cosmic microwave background is not the same in all directions—there are smudges and clumps here and there. Now these smudges would have been the result of very small quantum fluctuations in this centimeter-sized piece of universe, and they are then stretched to astronomical scales by the inflationary expansion. Over hundreds of millions of years these sparse densities coalesced into galaxies and stars and planets—indeed, we wouldn't be here if it were not for them and inflation.

Here is a case that blends all the best of science—a serendipitous discovery (cosmic microwave background radiation) that was the result of building a more sensitive instrument (the radio telescope) that gave rise to a wildly imaginative theory of the beginning of the universe (inflation) intended to solve some deep paradoxes in the accepted theory (Big Bang) that motivated experimentalists to make more sophisticated devices to gather more sensitive measurements that led to a theory of how galaxies and stars and our planet—and we—arose out of the primordial dot. And the circle isn't closed; theorists are still at work and Miller is still launching balloons, now looking for ephemeral gravitational waves—"if they're there", she says—another black cat in a dark room.

We have taken a quick tour of physics and cosmology using three researchers who exemplify different approaches to asking how the universe really is. Brian Greene is concerned with the deepest questions of how to describe a universe that we can't really imagine but may be able to solve mathematically; David Helfand sees solving some very hard but accessible problems by using the universe as his laboratory and thereby creating a long list of new questions; and Amber Miller wants to know about a moment in time that occurred so many billions of years ago that one could call it creation and that may contain a fundamental limit to what we can know of our universe.

This case history also brings up an important point about the uses of ignorance and also what it won't do. By emphasizing the questions, you have heard and understood, I hope, some of the most sophisticated issues of modern cosmology and physics. But can you solve them? No. You don't have the tools, the mathematics, the intuition, the technical expertise, to actually be a physicist. Science is indeed a technical and sometimes difficult activity that requires training and experience—lots of it. But is that the point? What is critical is not that everyone in America becomes a scientist. Rather it is that everyone can understand what's going on, what the stakes are, what the game is about. Science is not exclusionary; it does not belong to a small coterie of eggheads speaking a secret language. You can watch a sporting event and enjoy it without having the training or skills of an athlete. You can enjoy a painting or a symphony without possessing any of the know-how of an artist or a musician. Why not science? There is no more sense in getting hung up with the details of experimental results or systems of differential equations than there is with chord structures and harmonies in a musical composition.

3. THAT THING YOU THINK YOU THINK WITH

The smartest thing I've ever heard said about the brain was from the comic Emo Philips. "I always thought the brain was the most wonderful organ in my body; and then one day it occurred to me, 'Wait a minute, who's telling me that?'"

And that gets right to the heart of the matter, if you will pardon me that twisted metaphor.

Because you see, the single biggest problem with understanding the brain is having one. Not that it isn't smart enough. It isn't reliable. The first-person experience of having a brain is just not remotely similar to any third-person explanation of how it works. We are regularly fooled by our brains. Constructed as they were from evolutionary pressures directed at solving problems such as finding food before becoming food, they are ill equipped to solve problems like how they work. Just as quantum mechanical descriptions of the physical world are weirdly unintuitive to our brains, so biological and chemical explanations of the brain are even more weirdly unintuitive to itself. I mentioned in the last case history that a hard problem for astrophysicists and cosmologists is one of perspective—they are studying something, the universe, that they live inside of. You could say the same thing about your brain.

Take, for example, trying to figure out what are the most important questions to ask about the brain. In our terms here: where's the best ignorance? For more than 50 years the visual system has served as one of the premier model systems for brain research. The retina, a five-layered piece of brain tissue covering the inside of the back of your eyeball, has been dubbed a tiny brain, processing visual input in a complexly connected circuit of cells that manipulate the raw image that falls upon it from the outside world, before sending it along to higher centers in the brain for yet more processing, until a visual perception reaches your consciousness—and all in a flash of a few dozens of milliseconds. A batter in a professional baseball game has less than 400 milliseconds to make up his mind about whether to swing at a 3-inch diameter sphere traveling the 60 feet and 6 inches from the pitcher at 90 miles per hour. Since making the decision and the coordinated muscle movement of swinging the bat take up part of this time, the work of the visual system must be accomplished in something like 250 milliseconds. Pretty fancy stuff that can do that, no? Certainly worth figuring that out, no? This must be some high-class brain ignorance. Well, let's see.

Because of the apparent difficulty of visual tasks and the effortlessness with which we seem to accomplish them, the visual system has been regarded as one of the highest developments of evolution. Indeed, one of the most common arguments made against evolution, and one that even worried Darwin, is how something as marvelously complex as the eye could have developed in small steps by random mutations. (In fact, it appears that visual systems of varying sorts, some better than those found in mammals like us, have evolved as many as 10 different times in evolution—counterintuitively, it seems to be fairly easy to evolve eyes.) Being such visually oriented animals, we have understandably assumed that vision is a very high-order brain process and that studying vision will therefore tell us a lot about how the brain does all the other amazing stuff it does. There are so many neuroscientists working on the visual system that they have formed a subfield that has its own annual meeting to discuss current research. The Association for Research in Vision and Opthamology (ARVO) has more than 12,000 members. The eye is a perfect example of a model system—the retina is accessible, well-organized (meaning that it has a limited number of cell types that are connected to each other in stereotypical patterns called circuits—that is, one can make a wiring diagram of the retina much the way one could for a radio), and it performs a straightforward if complicated task. It can therefore be investigated for its own sake, and also because it will reveal fundamental principles about how the whole brain works. So far so good.

At what might be considered the opposite end of the scale is a task like walking—mundane, simple, something every healthy person over the age of 12 months does. It feels thoughtless, reflexive, unconscious, so we take it for granted because it seems to use so little of our brain power. Running to first base certainly appears far less neurologically demanding than getting the bat on the ball in the first place.

But it happens that we have become fairly adept at producing technology that mimics the visual system—photography, television, movies, pattern recognition algorithms. Doing what the visual system does can at least be imperfectly reproduced in our technology. Not so true for walking. More than a century of robotics research has failed to produce a machine that can walk more than a few steps on two legs, let alone go backward or up a slightly sloping plane, not to even mention steps. Walking around on two legs is in fact in many ways a more complex and demanding mental task than much of what goes on in the visual system. Daniel Wolpert, of Cambridge University, is fond of pointing out that IBM's Deep Blue supercomputer is capable of beating a grand master at the game of chess, but no computer has yet been developed that can move a chess piece from one square to another as well as a 3-year-old child.

So what's the more complex thing for the brain to do? Seeing or moving? Which would make the better model system for understanding how brains work? Have 12,000 neuroscientists been looking at the wrong model system? The short answer: Quite possibly.

The nervous system is often divided up into two main functional branches: sensory systems and motor systems. There are other ways to make distinctions between the various functions of the nervous system (conscious versus instinctive, for example), but this seems the most fundamental, and the sensory and motor divisions tend to divide neuroscientists as well. In much of neuroscience, you are either a sensory or a motor researcher.

The sensory systems include the five basic senses of vision, audition, touch, olfaction, and taste, although there are many more senses that could be included in this list, which hasn't been substantially revised since Aristotle first enunciated it. For example, just within touch there is pain (piercing and throbbing), temperature, itch, friction and rubbing, and hard and soft touch. Then there is the sixth sense, proprioception, a word that I still have to sound out to say, that simply means knowing the position of your body, and in particular of your head, at any moment. It is rarely listed as one of the main senses, but without it the world would jump around in a dizzying manner, you would be unable to stand or sit, let alone walk, and it is doubtful that you would even have a real sense of yourself in the world.

Motor systems refer to the parts of the brain that initiate and control action or behavior, that is, movement. Some of those movements can be the big ones like reaching for something, performing an athletic feat, or walking, and some of them can be quite small and unconscious like the tiny but constant movements of your eyes called saccades, or regular and repeatable movements like chewing or breathing.

Admit it though, even you find all that motor system stuff less appealing than the cool sensory systems that give us the perception of a beautiful painting, concert, perfume, or sumptuous meal, that allow us to appreciate a majestic landscape or a pretty face; plus they warn us of danger, keep us from bumping into things, and generally seem to make life more interesting. But the motor system—it even sounds like a boring machine, a kind of vocational rather than intellectual part of the nervous system—may in fact hold the key to our cognitively rich lives. This is why having a brain is such an obstacle to understanding how one works. Our opinions about what parts of it are more interesting or more complex are so terribly biased as to be nearly worthless—worse than worthless, they are obstacles.

Consider that one of the highest cognitive abilities we possess is language. Communicating ideas through speech is perhaps uniquely human and unquestionably has allowed us to develop and transmit all the important trappings of culture—art, history, philosophy, science. And this most cognitive of all brain activities is basically a motor act—we speak by controlling and coordinating a vast array of muscles in our chests, throats, tongues, and lips. Andre Breton, the leader of the Surrealist movement, if it can properly be said to have had a leader, once remarked that the speed of speech is faster than that of thought. Yes, we can all believe that, having let some utterance escape from our mouths that we are immediately sorry for. But aside from the humorous implications, given a moment's thought, it's clearly true. We don't think long, if at all, about the words we are saying in the middle of a conversation; they just "come out." All of this supposedly high-level cognitive apparatus in the brain ends up as an almost reflexive motor act.

So the brain and how it works may be the biggest question in biological science, but have we got the right small questions? From philosophers to undergraduates we all seem fascinated with the debate about whether it's even possible for a brain to understand itself. Throughout history it has always been compared to the most complicated technology of the day—it was all pneumatics and hydraulics to Aristotle and the Greeks and Romans of antiquity with their fabulous aqueducts and sewage systems; then it was a clockwork-like gadget when newly invented time-pieces full of springs and miniature levers got everyone to church on time and started humans on the miserable path to deadlines and scheduling; then it was a complex engine in the Industrial Revolution and more recently a computer; and today it is compared, predictably, to the Web. The two things that are common about all of these comparisons are that they recognize the brain as a very complicated thing, and they are all otherwise mechanistically incorrect. We still don't know how it works.

There's a lot that could be said about the brain and about the field of neuroscience, especially about what we don't know. Anything written about the brain is necessarily incomplete, and that is no less true for this case history in neuroscience. Thus, I have resorted to constructing a case history by piecing together the work of multiple neuroscientists. I have focused on three in particular who are trying to understand the brain by understanding, first of all, that our current knowledge, the result of decades of modern brain research, rich as it is, has a bias and may be sending us off in some mistaken directions. The neuroscientists in this selected group, who work with and represent a much larger contingent of colleagues who I am not mentioning specifically, are going back before some of our current conceptions arose, asking fundamental questions that were assumed to be settled. In other, simpler words, they are actually creating new areas of ignorance where we thought things were already known. And this is progress.

Larry Abbott is a theoretical neuroscientist. That is, he is a real person and a real scientist, but he asks questions about brain function by using computer-generated mathematical models of how bits and pieces of it might work. The value of this kind of thing is that he can ask questions for which there are currently no good experiments, for which the technology isn't available or there are ethical considerations. Computer-generated mathematical models can use a million neurons at a time, or more, while experimentally it is impossible to check the activity of each of even a few neurons. Models do this by using statistics and it gets frankly complicated mathematically, so it's best to leave this part to the professionals. But just to give you a flavor for how it can work, imagine that you wanted to figure out the air pressure in the room where you're sitting. You could use a model that described the average behavior of air molecules, and it would not be essential to know the actual activity—the position, speed, and direction—of each air molecule at each moment to know the room's air pressure. The average behavior of a very large number of individual particles can result in a very precise value. So it may be with the brain, where knowing the average activity of neurons under different circumstances could predict with great accuracy how the brain performs a task, even a complex one.

Theoretical work in biology is relatively new compared to physics, where it has a long and rather successful history. Indeed, Abbott, like many other theoretical neuroscientists, was trained initially as a physicist. The application of mathematics to biological problems has come slowly, but nowhere as rapidly or as importantly as in studies of the brain. This is not without resistance because the mathematical skills necessary to do this work are not generally part of the training of a typical biologist. A lot of what theorists do has the aura of cabalistic symbolism—multiple equations and odd symbols pile up to cover the pages in their journal articles. "Simplifying" assumptions are made that don't seem obvious, or simple. Among experimentalists there is often a suspicion that all this mumbo jumbo is just a bluff hiding a lack of data.

I made a comparison of neuroscience to quantum physics earlier in this case history, about how it is so unintuitive to our brains. To take that a step further, the difference between modern physics and brain science is that the unintuitive thoughts one has to think in physics can be done with the language of mathematics. We don't use math in biology that way—at least not yet. The objection to equations most often voiced by biologists is that they oversimplify, that you cannot capture the complexity of biology, of a biological system, whether it be a single cell or a whole animal, in an equation. Nonsense, I say. You can capture the whole physical universe in a few of them. This is another case of pre-Copernican thinking creeping into our reasoning as received wisdom. The brain seems complicated, so its explanation must be also.

I have to admit that I do not have the mathematical sophistication to follow most of the theoretical propositions being produced by these computational neurobiologists, as they have come to be called. But I do think they have particularly strong access to ignorance. This is at least partly because, like modern physicists, they can use mathematics to frame questions that are hard, maybe impossible, to ask linguistically. With mathematics as your language you don't end up with paradoxical-sounding descriptions like the Emo Philips's quip that opened this case history. When it comes to asking about the brain, theoreticians have the further advantage of not being hampered by the technical limitations that are often part of experimentation, and not being constrained by earlier findings—findings that were often the result not so much of having a good question but of having certain technical capabilities that produced certain kinds of data (see my earlier remarks about voltage spikes in the nervous system in chapter 2). Of course, there are technical advances in math as well—new propositions, new techniques, added calculating power, and their application to nervous systems has been occasionally successful—but still the theoretician is not as constrained as the experimentalist.

Abbott is very thoughtful about choosing "precisely where along the frontier of ignorance I want to work." Talking in my class, it's interesting to listen to him wrestle with this problem. He makes the point that for a theorist all this freedom is a challenge. Without the technical constraints it's much more difficult to define a limit. "I want to claim that I want to discover the roots of consciousness, but I happen to believe that if I did say that, I wouldn't get anywhere." "You know that if you are too risky in your research you'll get nothing done. Or you can play it safe and reap rewards for doing essentially the same thing over and over again," but that's really not getting much done and "you have to force yourself not to do that." You have to find the "stuff that pushes the edges for you," and to do that you have to be honest and say, "What can I personally tackle?" "Also you have to know the times you live in. Is there enough information for me to make progress here? When do you yourself say you're not going to be able to solve this?" "So you have to introspect and that's the good part. But you have to guess too, and you could guess wrong. There are no guarantees." I hope you can see the struggle here and recognize that this struggle goes on in parallel with the actual work that he does, and it is a constant struggle, back and forth, always trying to locate that sweet spot of ignorance. Let's see it in action.

Abbott has some "simple" questions. "This morning when I opened the refrigerator I noticed that there wasn't much orange juice left. This evening on the way home from work I remembered to stop at the grocery store to pick up some OJ. What happened in my brain that put that thought in there and retrieved it at the right time, 10 hours later?" The simplicity of this question, like that of understanding how we walk, can be deceiving. Appropriately, Abbott calls this the "memory of the commonplace," the vast number of things we remember that are not exceptional or are not practiced, but that nonetheless occupy memory space in our brains. This all seems so unremarkable, so easy to ignore, as to dismiss it as not being that important. It is so woven into our daily lives; we do it all day, every day, and even overnight. How easy it would be to overlook this pathetically obvious kind of neural activity as being too common to be significant. Yet when we think very hard about this question, it actually becomes harder to understand. In brain work especially, this is a clue that you may be on to something.

What is it about the orange juice example that is so tantalizing? For one there is a delay. You think of it once in the morning and then not again until 10 hours or more later, after a day during which your brain did a lot of other things—if you're a theoretical neuroscientist like Larry Abbott it did a lot of very complicated things. Yet there it is, sometimes tripped off by a cue of some sort, the grocery store comes into view, you see an ad for apples, the radio reports on Jews in some West Bank settlement, or any of a zillion things that may be strongly or weakly related to orange juice. And often there is no discernable, that is, conscious, cue at all, it just seems to be there. How does the brain keep that not-so-special memory alive for so long?

There is a similar kind of memory, called recognition memory that is noteworthy because we seem to have a remarkably huge capacity for it. There are now famous experiments in which subjects have been shown as many as 10,000 images in a relatively short time—just flashes of each one. Then when shown a second group of images and asked to identify which ones they had seen in the previous group, they were able to respond with an almost unbelievable accuracy of over 90%. If it weren't for the truly astonishing numbers involved, this wouldn't seem like taxing brain work—just watching some pictures flash by and recalling whether you've seen them before. You don't have to make a list of the scenes you've seen (that would be very hard); you don't have to describe the scenes (that would also be hard); you just have to identify them as familiar. But 10,000 of them, with better than 90% accuracy!

These and a few other similar considerations began to nag at Abbott. Were we missing something here? Because it didn't seem like our ideas about memories formed from synapses switching on and off could really describe this level of dynamic memory capacity—thousands of memories a day, some fleeting, some longer lasting, many not even consciously recorded, most soon forgotten, at least in their details. So you see that by asking the right question (and one that could have been asked 50 years ago just as easily), a window is cracked open ever so slightly. I almost hesitate to use the window-opening metaphor here because it suggests a beam of radiant light coming into the darkened room, when actually it is almost exactly the opposite—an apparently well-lit room is suddenly darkened by the stealthy entrance of an unimagined ignorance just outside the room. What we thought we knew so well can't be the whole story; it may not be any of the real story.

Around the early 2000s Stefano Fusi, another computational neuroscientist with a physics background, was working independently on this problem and he and Abbott independently came to a result that appeared to be catastrophic to understanding how memory works. Since 2005 they have joined forces and worked together, and Stefano was also a guest in my course on ignorance, although some years after Larry Abbott. The rest of this history interweaves both of their work.

The catastrophic nature of the problem was that we simply didn't have enough synapses to remember the things we do, all those commonplace memories, if all the current models for how memories persist in our minds were true. And the difference wasn't by a little bit, a decimal point or two of adjustment or a little tweak. It was, well, catastrophic. All the accepted models of memory formation relied on the notion that memories were composed of some number of synapses, the connections between the cells in our brains, whose strength had been modified, and that at a later time this network of active synapses could be accessed and would be perceived as a memory. The assumption was that, like computers, the more switches (synapses in the brain, transistors in the computer) you had, the more you could remember. This is called scalability. In a scalable system the process by which something is accomplished remains the same and if you want more of it, you just add more hardware (e.g., switches). The human brain has about a 100 trillion synapses (a number represented as 10^{14}, that is, 10 multiplied by itself 14 times). So even if a memory required a hundred synapses (10^2), there was enough hardware for 10^{12} memories—about a trillion memories, which certainly seems like more than enough.

But Fusi and Abbott (along with another neurophysicist hybrid named David Amit, who was Fusi's original mentor) had stumbled upon a dismaying problem with one of the assumptions of the accepted models. Without going into the technical details, they found that the number of memories in a wet, warm biological brain does not scale with the number of switches (synapses), as is the case with the transistors of a cold, hard computing machine. Instead, they only scale with the logarithm of the number of synapses. The logarithm is that number that tells you how many times to multiply by 10—in other words, the number of memories in a brain with 10^{14} synapses maxed out at 14, not a trillion. 14! That's what they mean by a catastrophe. (To be entirely accurate here, they actually scale with something called the natural logarithm, but that value only turns out to be about 36, not really much of an improvement.)

When Fusi first came upon this result, he was a young scientist, and a physicist no less, and couldn't get the result published in any major journal. An accidental meeting and conversation with Larry Abbott revealed that he also had stumbled upon a similar result but was suspicious of it, given the consequences. They began working together, and the result only became more and more robust. Still this discrepancy, shall we call it, was being largely ignored by the greater neuroscience community until finally Larry was asked to give a Plenary lecture at the Society for Neuroscience annual meeting in 2004 and presented his and Fusi's findings to a large captive audience.

The crucial thing about this finding is that it changes almost everything about the way we think the brain stores memories. Although Fusi and Abbott show this by performing careful calculations, you can get an intuitive appreciation for the problem—a system that learns quickly, as our brain does in commonplace mode, will also forget quickly. This is because new memories are constantly being formed that write over the old ones, so nothing lasts for long while the brain is active. Pretty quickly the synapses used in one memory begin getting incorporated into others and then there is less and less of the original memory left, and eventually it is no longer recognizable, that is, we have forgotten it. This also tells us that forgetting is due to the degradation of memory, not by time, as most of us are wont to think, but by ongoing activity. Forgetfulness, especially the kind that so worries people, like forgetting where you just put your keys or why you walked into a room, is not due so much to age as to overworking the poor contraption. New memories crowd out older ones, even on the time scale of minutes.

Of course, we have long-lasting memories, but that requires that the system making the memories does so only very slowly, too slowly to explain our normal experience of recognizing vast numbers of familiar things. Practiced or exceptional memory may work this way—long-term learning clearly requires slower processes and so it is more difficult. There is not only more than one kind of memory, but there must be more than one kind of mechanism for remembering.

Well, this story has a happy ending, at least for a book on ignorance, because the catastrophe remains unresolved. As Fusi says, "We were studying a problem, not a solution." Of course, there are some hypotheses, but they are mostly of the sort that we have to look at things we never looked at before, because the established principles do not contain the solution. The solution is in a new dark room with new black cats scurrying about—or not.

John Krakauer is a brash, young neuroscientist whose cultured demeanor and English accent barely cloak an unswerving scientific toughness. Krakauer is a medical doctor who slipped into research during his residency and has never left it. Using perhaps the polar opposite of the theoretical approach, he nonetheless has come to a similar conclusion that motor systems are the improbable key to understanding how the brain works. Two of his mantras are that "Plants don't have nervous systems, because they don't go anywhere" and "The reason to exist is to act." He now manages both a clinical practice and a research program. His quick, and often racy, sense of humor belies his deep thinking about what it is the brain is up to. Krakauer asks my class a simple question: "What muscle contracts first on pressing an elevator button?" This shouldn't be difficult. We all think back, running through a movie in our heads, our brains, of ourselves reaching up to push the button in an elevator. We've all done this hundreds, thousands of times, but the answer still shocks us: "The gastroc muscles," he says, "in your leg (this is one of the two long muscles of your calf) on the same side as the arm that you will lift to push the button. And if you didn't do this, tighten this muscle just slightly in anticipation of your fairly heavy arm (approximately 8 ½ lb!) being lifted like an extended lever, you would topple over." What is so shocking about this is that our brain has worked out the problem, not a trivial one in engineering terms by the way, made these anticipatory postural adjustments in muscles that are nowhere near the arm being lifted, and we have no access to the process. Imagine how much more complex it gets if you are carrying groceries in the other arm, but still you don't experience it as difficult.

This particular example reminds me of a pointed question asked by the philosopher Ludwig Wittgenstein. "When you raise your arm, your arm goes up. But what is left over if you subtract the fact that your arm goes up from the fact that you raise your arm?" There is the sense that in that tiny bit that's left over, for which we haven't got an exact name—intention or thought or decision—is a very important answer.

An even deeper question, which will at first sound silly (which is how deep questions often masquerade), is how come we all reach for the elevator button at about the same speed? Krakauer demonstrates this by reaching for his cup of "double shot, extra hot, skimmed latte" and asking why we all choose the same speed and motion when reaching for a cup of coffee. No one goes too slow or too fast, and if someone did you would think something was wrong with that person—mentally or physically. These striking regularities appear in most simple actions. Virtually everyone makes a fairly straight-line approach to the cup—not from above or below or a myriad other ways of reaching for the cup. Indeed, this is an enduring question in the neurobiology of action—what is called the "degrees of freedom problem"—that, given a nearly infinite number of ways to reach for the cup, why do we all choose the same one? "Why isn't there infinite procrastination as the brain tries to determine which of the many possible ways it will use?" "No one knows why," says young Dr. Krakauer. No one knows why.

Of course, there are all sorts of factors one could theorize about—maximizing efficiency, speed versus error, energy costs—and there are equations and graphs from dozens of experiments describing the effects of these and other factors on movement choice, but none yet fully explain this exquisite mystery. The medical doctor Krakauer points out, however, that these are important elements in designing rehabilitation programs for stroke patients or other victims of movement pathologies. What should we actually be working on to improve lost movement and coordination? What are the critical aspects of making movements that are lost due to pathology or injury?

Parkinson's patients are a particularly interesting example. The disease is marked outwardly by slowed movements, from walking to reaching for a cup of coffee. But, as Krakauer dryly notes, "They never get run over." If the problem were some sort of "execution deficiency"—poor muscle control or damaged communication between nerves and muscles—they would be likely to suffer many more accidental injuries than they do. Something seems amiss here, and this began a series of experiments with Parkinson's patients that pointed in a completely different direction from muscular dysfunction. A case where a simple question, why do Parkinson's patients move slowly, was at odds with the observed conditions (they don't have accidents), and getting the right question, not just more observations, was the key to understanding what was going on.

It's worth looking into this a bit because not only is the answer to this specific conundrum unexpected, it leads improbably to an understanding of the development of skill, for example, in sport, and of a new perspective on the whole of cognition. This is a wonderful example of how a little bit of the right ignorance can lead to undreamed-of insights in seemingly unrelated areas.

The answer, or at least the partial answer, is that Parkinson's patients, those in the earlier less debilitating stages of the disease, believe that they are moving at the correct speed; they are simply wrong about that. If you yell "Fire!" in a room full of these type of Parkinson's patients, they will beat you to the door. They can move just fine; they "choose" to go slowly. But why have they made this choice? You might think you could just ask them, but that won't work. They are unable to self-report on this. While they recognize they are moving slower than other people, and may even be embarrassed by it, they don't really know why they are moving more slowly than they could, so you can't just ask them. This is an example of why the brain is so poor an instrument for understanding how it works—at least through introspection. You can think about it all you want, and you will never get access to what your brain is doing computationally at any given moment. You only have access to a result, a behavior or a perception, that could have been reached in numerous indistinguishable ways. By the way, you are no more able to self-report why you have chosen the speed at which you walk or grasp for objects than a Parkinson's patient.

So what are the possible reasons that Parkinson's patients adopt the speed they do? It could be that they have miscalculated the speed-accuracy tradeoff—that is, how fast you can go without making serious errors, like falling down or knocking over the coffee. And, indeed, if you were to watch someone trying to navigate across an icy pond in slippery shoes, his or her gait and posture would look very Parkinsonian, due to the person's calculation of the risk of falling over.

Another possibility is that they have miscalculated the energy cost. They have an increased reluctance to move faster because some implicit calculation of the energy cost has gone awry. Think about yourself reaching for a glass of water on the table—you could do it much faster than you typically do and still not knock it over or spill it. Why do we "choose" to go more slowly than we are capable of? It could very well be that to beings from another planet we appear to move painfully slowly. So have we decided in some unconscious way that it's just not worth the effort to get to the glass any sooner?

In a variety of tests in which patients were asked to perform some action faster than they preferred, there was no difference in their accuracy than in healthy control subjects. In other words, accuracy was not the problem. Instead, Parkinson's patients appear to be making a faulty calculation about the cost to make movements of a certain speed, and this miscalculation causes them to move slower. They could go faster; they are simply not motivated to do so. It is this unintuitive insight that may bring together a variety of previously disparate facts, forming an explanation, or at least a model, of how movement control could be unexpectedly related to phenomena like addiction and skill. Stick with me on this; it's tricky but worth it.

The pathology of Parkinson's has been known for some time. A small collection of 25,000 neurons deep in the brain begins to die off, for unknown reasons. This is not many neurons (compared to the more than 80 billion in your brain), and it's surprising how their loss can cause such devastating and widespread effects. Some time ago it was discovered that these 25,000 neurons make connections widely throughout the brain and that they communicate with other neurons using the chemical dopamine. Dopamine is used by numerous other brain cells besides this group of 25,000 and appears to have a confusing array of effects on the brain—from movement control to addiction to reward circuits in learning, to schizophrenia.

The effect of dopamine on movement—at least on the slowed movement of Parkinson's patients—is really a question of reward calculations. Their diminished dopamine levels, caused by the loss of those 25,000 neurons, have compromised the brain's reward system. They are not motivated (note that the word *motivated* has the same root as *motor*) to move faster because for them there is no reward for doing so. The puzzling array of dopamine actions in the brain becomes clearer when you begin to think that behaviors are calculated by the brain to have a value associated with them—thus, movement and reward are not so discon-nected as you might have thought by simple introspection. This is surprising, and again unintuitive, because calculat-ing rewards seems to be a very cognitive activity—it involves planning, prediction, learned experience, intuition—"if I do this what are the chances of my being rewarded rather than disappointed or worse." The hot new "field" of neu-roeconomics, which attempts to understand our predilec-tions and aversions toward gambling and risk taking, uses calculations no different from the one the brain evolved to make about the cost-benefit of reaching for a piece of fruit at a particular speed. And thus dopamine is part of the motor system and the reward system at the same time—not by accident but because the motor and reward systems are highly connected in our prehistoric brains. Who would have thought?

I have outlined two small examples of work going on in neuroscience that are not necessarily about the big questions you might have thought dominated the field: the nature of consciousness, developmental pathologies of the nervous system, learning, and memory. They certainly touch on these issues, but the actual questions being asked are much more detailed. As important, they spring from almost obvi-ous questions—so obvious that they failed to attract much attention for decades. And the result is that a seemingly mature field finds a new direction—searching new dark rooms for black cats that no one is sure are going to be there. Just when it seemed that we neuroscientists were getting to the bottom of sensory systems and some explanations for how the brain works, a whole new area of ignorance opened up before us, with even greater promise. These new questions prove how full the magic well of the brain still is—full of questions still unimagined by the brain under study.

4. AUTOBIOGRAPHY

How does someone, a scientist, arrive at the questions that determine the course of the rest of his or her life? What is the personal debate that leads a young scientist to one field or another, to one set of unknowns, to a group of questions around which he or she will build a laboratory, a research group, a career, a life? We have seen that choosing the question is the essential act in science. It may then be surprising that this so often happens in what is described as a flash of insight, an "ah-ha" moment, an epiphany. But like many such experiences, careful reflection shows that there was, in fact, a long, if clandestine, preparation for the moment of recognition. Can we recover that process? Is there value in knowing the history? Can we understand the process better than just chalking it up to a happy accident? In retrospect, it is often possible to see the route that may not have been very apparent during the journey itself, that with perspective something sensible can be extracted from an otherwise haphazard appearing odyssey. There is of course the danger that "in retrospect" we distill what was in fact a chaotic process into a tidy linear narrative, but such is the nature of memory and the desire to have a story.

Here then I have decided to use my personal story, my own case history. It has some unusual elements in it, but they serve to highlight many of the decisive factors, both intended and serendipitous, common to the creation stories of scientific careers. And there is certainly more than enough ignorance woven through it to provide a lesson or two.

So here it is: Stuart Firestein, Professor of Neuroscience at the Department of Biological Sciences, Columbia University, New York.

I came to science late, after a career in, of all things, the theater, where I spent more than 15 years working professionally as a stage manager and director and had the opportunity to run my own repertory company. Although now there are university programs for training in the theater, the traditional path then was to apprentice yourself to professionals and learn by hanging around with accomplished artists. You started as a stage worker, setting up scenery and lighting equipment, attending production planning meetings, working the lights, or moving scenery at performances in the evening. You then graduated, if you desired and were any good, to assistant stage manager, an especially interesting position because you attended all the rehearsals, but the position had no set of assigned responsibilities. You simply did whatever came up, from getting coffee to copying scripts to organizing rehearsal props, to whatever. The nice thing about this position was that you spent a lot of time at the rehearsals, involved in the production but not burdened with responsibilities, and so you had the opportunity to see how things work—and just as important how they don't work. You could watch actors work out scenes and develop character business while directors tried to find the right staging and develop an ensemble feel. You were part of the production but also had a critical distance that allowed you to learn while doing. It's a system, I realize, not so different from the way we train graduate students in our laboratories, and it is one that I recommend highly. It is not so easy to create the condition where you can have perspective and involvement simultaneously, where you can be invested but not fully responsible, immersed yet without the pressures of liability. But having such an environment seems fundamental to a mentoring process that allows us to explore questions. This is one lesson from this case history.

I was pursuing a reasonably successful career as a theatrical director involved in a variety of productions from avant garde experimental works to frankly commercial projects, mainly with repertory companies up and down the Eastern seaboard. I often joked that my hometown in those days was the Amtrak Northeast Corridor. In early 1979 an opportunity came my way, which I took on a lark and probably because I needed the money, to move to San Francisco with a touring production. Somewhat to my surprise I found the active theater scene there attractive, and I decided to stay on and explore it. A few years later I was involved in a successful production with a promised long run, leaving my days relatively free. I had a lifelong interest in animal behavior that I had never really followed beyond popular reading, and this unusual moment of stable employment gave me an opportunity to pursue it more seriously. I decided to take a course at the local college, San Francisco State University, where I found a class in Animal Communication taught by a Professor named Hal Markowitz. I mention his name because it was a very happy accident that I ran into him. He became an important mentor to me, and mentors are often a critical part of the story for any scientist. I had never been to college before and here I was a 30-year-old student. I found it remarkable. Someone stands up in front of a group of people and tells you everything he/she knows about something. What a great idea, who thought of this? As it turns out, I think it was Aristotle.

Hal Markowitz, my latter-day Aristotle, was a generous fellow with his accumulated knowledge but also with his accumulated cynicism. The field of Animal Behavior can attract a lot of people who love animals and have deep relations with pets and other creatures they volunteer to work with. They are often delightful people with very good hearts, but they are rarely scientists. Hal Markowitz would have none of that; animal behavior was as serious a science as physics for Hal, and no less rigorous. I was impressed by this because, frankly, I was one of those folks who were interested in animal behavior because I liked animals. Not that Hal didn't like animals; he liked them in what I believe is a truer way than most—he liked animals for what they are, not how similar to humans they might be or what use they might be. Hal was not interested in whether animal thinking approached that of humans; he was interested in animal thinking, period. He wanted to know about all sorts of animals and how they behaved because the very variety of biology was for him the source of endless questions. The intensity of Hal's inquiry into something that had been a kind of hobby for me was something quite new, quite striking. I had not imagined how much there could be to know about animal behavior, how many deep questions there could be, if you just refused to be satisfied with a superficial or cursory exploration. This was an adventure.

Hal posed questions that were hard. Does a dog urinating on a tree intend to communicate something? Does it appreciate the message that the next dog will sniff? How about the message a deer rubbing on a tree communicates to a predator? What's different about those? Is it better to study animal behavior in the wild, like Konrad Lorenz and the European school of ethology, or in the laboratory like B.F. Skinner and the American discipline of behaviorism? Philosophically one answers the question in favor of the ethologists, but the overwhelming explanatory power of the behaviorists cannot be dismissed.

Hal hooked me. I took another course from him and we became friends (still are). He convinced me to take more biology courses and to consider working toward a degree. I was 30 years old, I had been working in the theater since I was 18, had never been to college, and had never thought of earning a degree. Could I really handle serious chemistry and physics and math? Hal assured me so, although I'm not so sure to this day that he was all that convinced himself. I enrolled as a full-time undergraduate student, but I kept up some work in the theater, mostly tech work at night, just in case. Organic Chemistry was a Biology degree requirement, and I realized that would be the great challenge. I would have to defy *Orgo*, as it is known among students, the great monster that sorts out the real science students from the wannabes. Or so is its reputation. If I couldn't get through Orgo, then I would know this was simply beyond me. So I took it as soon as I could. To no one's surprise more than my own, it turned out to be my favorite course and one of my best. At least part of the reason was that organic chemistry requires a lot of memorization, and this is where it gets its fearful reputation as an impossible course. The thing was, memorization was trivial for me. For 15 or so years I had been memorizing scripts for a living. You often don't know what you bring to the table, and this is the second lesson from this case history.

Once the memorization was not a factor, the complexity of carbon-based chemistry, one of the most beautiful rule-based systems in the universe, was well...fun. I aced Orgo, and it seemed I was on my way. To where? After 4 years I had a BS in biology. I was proud of the accomplishment, but truth be told, the degree was worthless as far as altering my life. A career in biology would require a higher degree, a PhD earned in graduate school. I applied to a few graduate programs and thought that I would leave it to the admissions committees to decide my fate. If I was accepted into a graduate program, I would give up the theater for good and change my career to science; if not, I would go back to the theater full time, with my BS, my pride, and a curiously interesting background for a theatrical director. Remarkable, in retrospect, that I left such a life-altering decision up to an anonymous admissions committee, but I have since learned that this is often the case. Scientists are a strange lot this way—they seek to control every aspect of an experiment, but life decisions they leave up to committees and review panels with often anonymous memberships.

As it happens the Neuroscience program at the University of California, Berkeley called to say that they had accepted me into their program. Surely the result of a clerical error, I thought. No matter, I accepted their offer immediately, and at 35 years of age, I quit the theater for good and showed up for graduate school.

I wish I had a sensible narrative about all this, but as you can see it was largely happenstance—some good luck, an excellent mentor, the right bit of preparation, some more good luck, and perhaps a clerical error. Will Rogers used to say that people don't so much fall in love as step in it. I think the same may often be said of science. Even those who know from their third birthday that they will be a scientist can't tell you precisely how they got to be doing exactly what they are doing. They try this or that, run into a professor or a graduate student who takes him or her under their wing and infects them with their mystery, and that's it.

Only in retrospect does it seem that the question and you were made for each other. This is the fallacy of design, not much different from the misleading arguments made against evolution with its random mutations and post hoc selection. Once the function of something is known, it always appears to have been designed. This, of course, was Darwin's great intellectual leap—to see that such utilitarian structures as eyes were not designed for their purpose, but that their purpose selected for them. We often wonder at the miraculous circumstances that bring together two lovers. How in the more than 7 billion people inhabiting the planet did these two people, so ideally suited to each other, find one another? What are the chances? Actually they are probably pretty good, which is why it happens so regularly. For one we are wrong to believe that there is only one other person in the world who is "perfect" for each of us. Probably there are thousands. And then we often start with less than perfect, and each becomes more perfect—or we get divorced. Is this the case with scientific questions? There certainly is no lack of questions, I hope we've established that. So you run into them, you can hardly help running into them, and then precipitously one of them sticks, for reasons that may be unfathomable, and in the end may not matter. One of them hooks you because the bait looks especially tasty, or you are especially hungry. And then sometimes it doesn't last and you get divorced. This is the third, or is it the fourth, lesson from this ever more bizarre case history. There's a lot to be said for making the most of happy accidents, and relying on happy accidents is no shame. But keep in mind that "Chance favors the prepared mind," as Louis Pasteur famously noted.

I am often asked if I miss the theater, by which I guess people mean the excitement, the glamour, the creativity. The short answer is, no. The glamour I don't know much about; I remember hard work, late nights, exhaustion, fear, arguments, people crying, but only fleeting moments of what might be called glamour. So nothing to miss there. But as for excitement and creativity, I don't think that science has any less of that than the theater, or any of the arts. I knew actors who showed up on opening night with the same performance they brought to the start of rehearsals 6 weeks earlier—which was often not much different from the role they had been performing in everything else they did. I knew directors who used the same bag of tricks in production after production; in fact, they were often hired by producers for their reliable bag of tricks. This is not what I consider creative. Of course, I know scientists who are the same, whose work is just as pedestrian and derivative and repetitive. But then there are the creative ones, just like there are creative artists, and they are no less adventurous, no less bold, no less perceptive than the best of artists.

And the excitement—I am afraid that it is impossible to convey completely the excitement of discovery, of seeing the result of an experiment and knowing that you know something new, something fundamental, and that for this moment at least, only you, in the entire world, knows it. When I was a graduate student, I was working late one night at the lab and I obtained a really unexpected result that answered a long-standing question. It was quite late and there was no one to tell. I remember going home that night and thinking that I should be extra careful in traffic because only I knew this thing and I needed to protect it. There was a kind of thrill in this and the whole world looked different that night.

The one rational decision I did make was that in graduate school I would work on something more reductionist than behavior. The brain is the source of behavior, so I decided to look at how the brain works. This was perhaps a little naïve, but naiveté can be important at certain times, like at the beginning. As Hal Markowitz used to put it, all behavior is just stretching and squirting—neurons squirt out neurotransmitters that cause muscles to stretch. So I thought, let's study how neurons squirt neurotransmitters, and this may lead to a deeper, or at least more mechanistic, understanding of behavior. Admittedly this may sound radically reductionist, and I am perhaps overstating it a bit. But there is some middle ground between simply observing animals or humans behave, and trying to figure out what's going on inside of them that makes them behave that particular way, and I thought, no I knew, that this was the ground for me. I hit upon the olfactory system, the sense of smell, as a possible place where that middle ground may be accessible to study. Smell governs, or at least modulates, a wide range of behaviors in many animals, including those associated with feeding, aggression, sex—pretty much all the things that matter. I thought that perhaps learning about how smell works would eventually lead me back to behavior, that if I became expert in the physiology of smell and odor perception, then I could go back to behavioral studies with a new and deeper appreciation for its underlying causes. About all of this I was almost completely wrong, but about the study of smell being a frontier I was presciently correct. Lesson five (four?), predictions are useless, except for when they are helpful.

I joined the laboratory of Professor Frank Werblin, my second great piece of luck in the mentor category. Frank's lab worked on the retina—remember, that small piece of brain tissue that coats the back of the eyeball. In deference to Frank, I agreed to try working in the retina, at least to learn some techniques that could be applied to olfaction later on. This turned out to be a disaster. I got nowhere. As mentioned in the previous case study, the retina is a complicated but well-studied bit of brain. An excellent model system—the retina has been compared to a little brain in its complexity and function—it has five different types of brain cells that are connected to each other in a sophisticated network of circuits. The retina not only receives light rays, it operates on them in ways that produce a coherent input to the brain. But for all that, it failed to excite me. It seemed that all the big questions about how the retina worked had been answered and what was left were the details. These are important details to be sure, and, indeed, the retina remains a vigorous area of neuroscience research. But for me it was not the dark room that I wanted to venture into.

Frank Werblin was a generous mentor and a true scientist. Working nights and weekends, I produced some data from olfactory neurons and when I showed the results to Frank he insisted, after recovering from the fact that my data came from the nose not the eye, that olfaction was where I should work. This was not an easy decision for him—his funding was for work in the retina, his excellent reputation had been made in the retina, his colleagues were all in the retina field. But Frank believed in data and in being guided by questions that mattered to you. I think he also liked being a little bit outside the normal distribution, and olfaction was certainly that.

So I went to work on the olfactory system within his lab. This was a true stroke of good fortune because research on the retina and visual system was far more advanced than the olfactory system. I thus had my "apprenticeship" among students and postdocs in the more sophisticated vision field. I was held to higher standards than if I had just started right out in olfaction. This turned out to be quite important, itself another lesson: you always have to find the highest current standard and measure your work against that. I am to this day indebted not only to Frank but to the graduate students and postdoctoral fellows in his lab who relentlessly challenged me and generously aided me, forced me to work harder and then made that work more productive. Regardless of what the mythology may say, science is rarely done in isolation.

I managed to complete my PhD work just as I turned 40, something of a milestone as you might imagine. I was offered a postdoctoral position by Dr. Gordon Shepherd at Yale University's Medical School. Gordon Shepherd is one of the truly decent people in science. I have never seen him put his own self-interest above those of the students in his laboratory. He is mild mannered, decorous, and decent, almost beyond belief. But I have also seen him fight like hell over a scientific issue or a paper that was incorrectly handled by an editor or reviewer.

I was attracted to Gordon's laboratory because he saw olfaction as part of mainstream neuroscience. Not everyone did. Olfaction was a bit of a neuroscience cul-de-sac in those days. As a sensory system, it was thought to be idiosyncratic, somehow unique in the way it worked and therefore difficult to make headway. It was, you might say, the opposite of a model system. Gordon saw it differently. Trained at the National Institutes of Health in synaptic physiology and basic neurophysiology, Gordon eschewed the specialness of the olfactory system and instead believed that olfaction should and did obey the rules of neuroscience, known and unknown, just like every other brain system. This was critical because it meant that the advances being made in vision and hearing and touch and in other brain functions could be relevant to understanding olfaction if applied thoughtfully. As important, it meant that things we learned in olfaction could also be relevant to understanding other parts of the brain, and that gave one the feeling of both belonging and contributing. Olfaction in Gordon's lab was not an isolated island of neuroscience.

It wasn't a sure bet in those days—the sense of smell was indeed puzzling, still is in many ways. How could you discriminate the many thousands of chemicals that were known to be odors? Why does one chemical compound smell like caraway and another almost identical compound smell like spearmint? How can adding a single carbon atom to a molecule change its smell from parmesan cheese to rancid sweat? How do smells evoke vivid memories that are decades old? Many of these questions remain current or have morphed into newer versions that are more sophisticated. It is not my intention here to survey the ignorance of olfaction, which would be, like any field, worthy of its own chapter. I mention these questions to show that the field was at the time wide open, full of ignorance and fallow. I was lucky to have stepped in it. I was smart to have stayed.

I have had such good fortune with mentors that I am always surprised and a bit puzzled when I hear other people's horror stories. And they are legion: graduate students who have been taken advantage of, mistreated, forced to work on questions they found uninteresting or unimportant, uncredited for work they do or discoveries they made, ripped off, pissed off, crapped on. How could it go so horribly wrong when it seemed so happily right for me? Was I simply lucky? But three in a row? Was it because I was older and perhaps more mature and had different expectations? I don't know the answers, and I am disappointed to admit that there doesn't seem to be a formula that can be followed for training graduate students. I wish I knew the prescription so that I could be sure that I would be the mentor to my students that Hal and Frank and Gordon were to me. But there you have it, as much ignorance as there may be about the brain, there is also ignorance about how to study the brain and even how to prepare to study the brain.

. . .

I hope these four case histories have provided you with a feeling for the nuts and bolts of ignorance, the day-to-day battle that goes on in scientific laboratories and scientific minds with questions that range from the fundamental to the methodological, and that initiate and sustain scientific careers. They are merely examples of how the scientific enterprise is carried on by thousands of individual scientists in many hundreds of laboratories and institutes throughout the world, an enterprise that has been continuously pursued through nearly 15 generations. Its worldview is not one that has taken hold in all cultures, and the impetus to see the world as a tractable mystery is not one that is really common. Most human cultures have been dominated by nonscientific explanations, including our own until a mere few hundred years ago. Many still are.

We often use the word *ignorance* to denote a primitive or foolish set of beliefs. In fact, I would say that "explanation" is often primitive or foolish, and the recognition of ignorance is the beginning of scientific discourse. When we admit that something is unknown and inexplicable, then we admit also that it is worthy of investigation. David Helfand, the astronomer, traces how our view of the wind evolved from the primitive to the scientific: first "the wind is angry," followed by "the wind god is angry," and finally "the wind is a measurable form of energy." The first two statements provide a complete explanation but are clearly ignorant; the third shows our ignorance (we can't predict or alter the weather yet) but is surely less ignorant. Explanation rather than ignorance is the hallmark of intellectual narrowness.

Getting comfortable with ignorance is how a student becomes a scientist. How unfortunate that this transition is not available to the public at large, who are then left with the textbook view of science. While scientists use ignorance, consciously or unconsciously, in their daily activity, thinking about science from the perspective of ignorance can have an impact beyond the laboratory as well. Let me, in a final brief chapter, suggest how ignorance can also be useful in two areas of current concern and debate: public scientific literacy and education. My intention is only to make a few remarks on each of these topics in the hopes that the reader will be inspired to use the ignorance perspective to extend his or her own thoughts about these very public issues.

EIGHT

Coda

If you cannot—in the long run—tell everyone what you have been doing, your doing is worthless.

—Erwin Schrodinger, from "Science and Humanism, Physics in our Time," a lecture delivered in Dublin, 1950

PUBLIC AWARENESS OF SCIENCE

Science, more than ever, uses and requires public money. Scientists therefore have both a responsibility and, quite frankly, a necessity to educate the public, to engage them in the scientific enterprise. The beginning of Western science is often taken as the publication of Galileo's *Dialogue Concerning the Two Chief World Systems* in the late Renaissance. Notoriously Galileo got in some serious trouble with the Church powers over this work due, we are taught, to its heretical propositions about the universe, or what were then still called the heavens. In fact, it was not so much what Galileo said about the relation of the sun and the earth in his famous work; the Church fathers are believed to have mostly agreed with it, being intellectuals themselves, but they just hadn't worked out how to tell the literal Bible-believing public about it. The real objection was that Galileo, following the trend of the Renaissance occurring all around him, published this seminal work in Italian. It was the first book of science ever to be published in a vernacular language rather than in classical Latin or Greek, knowledge of which was restricted to a small class of intellectuals. It was not the ideas, heretical though they were, but rather their potentially wide dissemination that so worried the Church fathers.

And those churchmen were correct because Galileo's landmark work began a tradition of publishing science in common languages—Descartes in French, Hooke in English, Leibniz in German, and so forth. The public's direct experience of the empirical methods of science is widely regarded as responsible for the cultural transformation from the magical and mystical thinking that marked Western medieval thought, to the rationality of modern discourse. Indeed, public accessibility to science may have been the most important contribution of the Renaissance to scientific progress—even more, some might say, than all the remarkable findings of the period beginning with Galileo's book in 1652. By the time of Maxwell, Faraday, and Hooke, for example, the public's appetite for science was voracious. Science demonstrations were put on as entertainments in performance halls, and science books sold as briskly as novels.

Today, however, we find ourselves in a situation where science is as inaccessible to the public as if it were written in classical Latin. The citizenry is largely cut off from the primary activity of science and at best gets secondhand translations from an interposed media. Remarkable new findings are trumpeted in the press, but how they came about, what they may mean beyond a cure or new recreational technology, is rarely part of the story. The result is that the public rightly sees science as a huge fact book, an insurmountable mountain of information recorded in a virtually secret language.

It's no small matter for the citizenry to be able to participate in science and understand how their lives are being changed by it. For some reason it seems easier to access the artistic side of the culture, while the science part is daunting. But science and empirical thinking are as indelibly a part of Western culture as the arts and humanities. Maybe more so. Precisely where and how science started, whether with the Greeks or the Arabs, the Phoenicians or the early Asians, it has flowered in the West as nowhere else. For the 15 generations since Galileo, science has molded our thinking and altered our worldview, from how we think the solar system is organized to how we communicate over this nebulous but ubiquitous thing we so appropriately call "the Web." This brand of science has spread to other cultures and made itself into a global venture long before the word *globalization* was popularized. For better or worse, our world has been transformed in record time and to a degree unimaginable at the beginning of it all some 400 hundred years ago. And now you live in that world. Your children grow up in that world. You rely on that world. You should know about that world.

Another no less compelling reason to be in the know about science is that loads of your money, in tax dollars and corporate spending, are going to support it. US government support of scientific research and education is nearly 3.0% of the gross domestic product—to be more blunt about it, that's some $420 billion annually. Corporate research budgets account for two-thirds more than government spending, amounting to an additional $700 billion. Corporate research is reflected in the price you pay for energy, for drugs, for just about everything and anything. Admittedly, these numbers include military research (although only the nonclassified part of it), but it's all science no matter what its intended purpose, and it's all being billed to you.

Then there are all those thorny ethical issues that keep bubbling up from science—stem cell research, end-of-life definitions, health care expenses, nuclear power, climate change, biotech agriculture, genetic testing—and this list promises to continue growing in the future.

Clearly what we need is a crash course in *citizen science*—a way to humanize science so that it can be both appreciated and judged by an informed citizenry. Aggregating facts is useless if you don't have a context to interpret them, and this is even true for most scientists when faced with information outside their particular field of expertise. I'm a neurobiologist, but I don't know much more about quantum physics than the average musician, and I could no sooner read a physics paper in a science journal than I could read the score of a Brahms symphony. I'm an outsider, too. I feel your pain.

I believe this can be changed by introducing into the public discourse explanations of science that emphasize the unknown. Puzzles engage us, questions are more accessible than answers, and perhaps most important, emphasizing ignorance makes everyone feel more equal, the way the infinity of space pares everyone down to size. Journalists can aid in this cause, but scientists themselves must take the lead. They have to learn to talk in public about what they don't know without feeling this is an admission of stupidity. In science, dumb and ignorant are not the same. We all know this; it's how we talk to each other and to our graduate students. Can we also let the public in on the secret?

EDUCATION

> But so soon as I had achieved the entire course of study at the close of which one is usually received into the ranks of the learned, I entirely changed my opinion. For I found myself embarrassed with so many doubts and errors that it seemed to me that the effort to instruct myself had no effect other than the increasing discovery of my own ignorance.
>
> —Rene Descartes, *Discourse on the Method of Rightly Conducting the Reason and Seeking the Truth in the Sciences*, 1637

Perhaps the most important application of ignorance is in the sphere of education, particularly of scientists. Indeed I first saw the essential value of ignorance through teaching a course that failed to acknowledge it. The glazed-over eyes of students dutifully taking notes and highlighting line after line in a text of nearly 1,500 pages, the desperation to memorize facts for a test, the hand raised in the middle of a lecture to ask only, "Will that be on the exam?" These are all the symptoms of a failing educational strategy.

We must ask ourselves how we should educate scientists in the age of Google and whatever will supersede it. When all the facts are available with a few clicks, and probably in the not very distant future by simply asking the wall, or the television, or the cloud—wherever it is the computer is hidden—then teaching these facts will not be of much use. The business model of our Universities, in place now for nearly a thousand years, will need to be revised.

In a prescient and remarkable document from 1949 on "The German Universities" appear the following lines from a report by the Commission for University Reform in Germany:

Each lecturer in a technical university should possess the following abilities:

(a) To see beyond the limits of his subject matter. In his teaching to make the students aware of these limits, and to show them that beyond these limits forces come into play which are no longer entirely rational, but arise out of life and human society itself.
(b) To show in every subject the way that leads beyond its own narrow confines to broader horizons of its own.

What an extraordinary prescription, improbably coming from a joint government-academic commission no less. It is a clarion call from a half century ago for us to rethink the education of scientists. Yet to be implemented, we would do well to heed it now.

Instead of a system where the collection of facts is an end, where knowledge is equated with accumulation, where ignorance is rarely discussed, we will have to provide the Wiki-raised student with a taste of and for the boundaries, the edge of the widening circle of ignorance, how the data, which are not unimportant, frames the unknown. We must teach students how to think in questions, how to manage ignorance. W. B. Yeats admonished that "education is not the filling of a pail, but the lighting of a fire." Indeed. Time to get out the matches.

. . .

We are all scientists: trying to understand our environment, to make sense of input that is not always complete or sensible, looking for black cats in dark rooms. Our minds do their best to decipher a complex world with information gathered by our limited sensory organs. The process is familiar to us all. We occasionally do "experiments," testing this or that to see how closely it fits our theory of the world. But let's face it: we are mostly stumbling about in the dark. The occasional glimpse of genuine reality only confirms for us the extent of the darkness we live in, the scope of our ignorance. But why fight it? Why not enjoy the mystery of it all? After all, there's nothing a like a good puzzle, and it turns out, in this life, it's not hard to find one.

Notes

I have not included extensive notes and citations in order to make the reading experience smoother. There are a few places in the text where I could have gone on longer, but there always are, so I resisted as much as possible. I have here included a few extra notes and, where it was not clear in the text, the source of some material. I have also included an annotated reading list, including books and articles that I use in my class and others that I used to write this book. The articles are all available for download on the Ignorance Web site (http://ignorance.biology.columbia.edu) as PDF files; the books are almost all still in print or easily available from used book sites on the Web. My remarks about them should be considered personal opinions relevant specifically to this book and topic, and not complete reviews.

PAGE 2

Andrew Wiles description of searching in dark rooms is from a *Nova* interview on his publishing a solution to Fermat's Last Theorem in 2000.

PAGE 7

Pascal once said, by way of apology, at the end of a long note written to a friend, "I would have been briefer if I'd had more time." This has been credited at one time or another to Voltaire, Abraham Lincoln, Mark Twain, T. S. Eliot, and others. But the earliest and most credible source I could find was as follows:

> Blaise Pascal, *Lettres provinciales*, 1656–7, Number 16, this one written December 4, 1656. "Je n'ai fait/cette lettre-ci/plus longue que parce que je n'ai pas eu le loisir de la faire plus courte."

This is from a discussion thread on the Web site for the NPR program *A Way with Words*.

PAGE 11

New data in a continuation of this Berkeley study of information shows a million-fold increase in capacity and transmission. The Web site for the study group is as follows:

> http://www2.sims.berkeley.edu/research/projects/how-much-info/

You can find summaries of the report and the full report as a PDF there.

Also an article in *Science* online in 2010 reviews the data. That citation is as follows:

> Martin Hilbert and Priscila López, The world's technological capacity to store, communicate, and compute information. *Science, 332*(6025), 60–65. doi: 10.1126/science.1200970

A PDF of that report can be found at the Ignorance Web site.
Other relevant URLs:

> http://hmi.ucsd.edu/howmuchinfo_research_report_consum.php
> http://hmi.ucsd.edu/howmuchinfo.php

PAGE 14

There is considerable controversy over the rate of growth of scientific literature and how to measure the actual number of scientific publications. While this may sound like a bean-counter problem, in fact there is quite a bit at stake, including things like obtaining grants and judgments about promotions and tenure. Are all articles to be counted equally when it is clear that some are more important than others, or more extensive, or have a longer shelf life? Do we only count articles, or do conference presentations also matter, or online publications? Does an article have to be available in English to be counted? How often is an article cited by other scientists? So you can see it's a morass, but the outlines and general estimates are reasonably clear. If you want to delve into this further, I recommend a wonderful little book called *Big Science, Little Science* by Derek J. de Solla Price (Columbia University Press, 1961). I believe it is out of print and I was unable to locate any more than excerpts online, but a more dogged search might come up with a used copy or access through the Yale Library (de Solla Price was a Yale professor). De Solla Price was one of the first to apply serious quantitative methods to studying the scientific literature. This book, like many of his papers, is nonetheless very readable and accessible. He died in 1983, an unfortunately young man of 61, probably at the height of his powers.

This article is a recent updating of de Solla Price's work, and it is available on the Web and at the Ignorance Web site: Peder Olesen Larsen and Markus von Ins, The rate of growth in scientific publication and the decline in coverage provided by Science Citation Index. *Scientometrics*, *84*(3), 575–603 (2010).

PAGE 17

Erwin Schrödinger, one of the great philosopher-scientists, says that "In an honest search for knowledge you quite often have to abide by ignorance for an indefinite period." Erwin Schrödinger, *Nature and the Greeks and, Science and Humanism*. Cambridge, England: Cambridge University Press (1996). Reprinted in Canto series with a foreword by Roger Penrose, in 1996. These are written versions of two lectures, The Shearman Lectures, delivered by Schrödinger at University College London in May 1948. They are full of exciting ideas and fascinating historical perspective from a man who stood as close to the horizon as anyone.

PAGE 20

Mary Poovey recently wrote a noteworthy book titled *A History of the Modern Fact* in which she traces the development of the fact up to its current, perhaps overexalted position. A very readable account of something you might otherwise not think very much about—after all a fact is a fact, no? No. Mary Poovey, *A History of the Modern Fact: Problems of knowledge in the sciences of wealth and society*. Chicago: University of Chicago Press, 1998.

PAGE 32

J. B. S. Haldane, known for his keen and precise insights, admonished that "not only is the universe queerer than we suppose, it is queerer than we *can* suppose." The full quote from Haldane is:

> "I have no doubt that in reality the future will be vastly more surprising than anything I can imagine. Now my own suspicion is that the Universe is not only queerer than we suppose, but queerer than we *can* suppose."

J. B. S. Haldane, *Possible Worlds and Other Essays*. London: Chatto and Windus (1927), p. 286. This quote, or one very similar to it, is often credited to the astronomer Sir Arthur Eddington, but this seems to be a misattribution because there is no citation or record for Eddington ever having said it.

PAGE 33

In a similar vein, Nicholas Rescher, a philosopher and historian of science, has coined the term "Copernican cognitivism." Nicholas Rescher is one of the leading philosophers of science in contemporary studies with a remarkable record of productivity over six decades, including some 100 books (I didn't actually count them). I am always surprised to find that his books tend to be well known among academics but not widely known outside of philosophy. This is a pity because they are immensely readable. I have relied on many of his writings to spur my thinking and, happily (for me), found that we were in agreement on many things. Here are just a few of his books that I found to be especially rewarding:

Finitude: A Study of Cognitive Limits and Limitations. Heusenstamm, Germany: Ontos Verlag, 2010.

The Limits of Science. Pittsburgh: University of Pittsburgh Press, 1999 (1984).

Pluralism: Against the Demand for Consensus. Oxford, England: Oxford University Press, 1993.

Unknowability. New York: Lexington Books, 2009.

PAGE 33

In Edwin Abbott's 19th-century fantasy novel, a civilization called *Flatland* is populated by geometric beings (squares, circles, triangles)... This book is available from a gazillion publishers; it has apparently never been out of print since its first publication in 1884. I personally liked this version with lots of interesting annotations by Ian Stewart who also wrote an "updated" version of Flatland called Flatterland, which was unfortunately not quite as charming as the original. There are several animated cartoons based on the book as well, but I have found all of them to be stupid and boring compared to the book. None of them are in 3D yet—but then why would they be?

Ian Stewart, *The Annotated Flatland: A romance of many dimensions*. New York: Basic Books, 2008, p. xvii.

PAGE 40

As Rebecca Goldstein recounts in her excellent, and highly detailed, book on Gödel, his shyness and reluctance...

Rebecca Goldstein, *Incompleteness: The proof and paradox of Kurt Gödel*. New York: W.W. Norton & Company, 2005.

PAGE 53

Seymour L. Chapin, A legendary bon mot? Franklin's "What is the good of a newborn baby?" *Proceedings of the American Philosophical Society*, *129*(3), 278–290, September 1985, p. 278.

I find it remarkable that in a country that has been the world leader in scientific research and that has gotten so much economic growth and general well-being from the fruits of that research that we have to be reminded of how important it is to take the long view. We are busy quoting the Founding Fathers for so many things of dubious value that Franklin's statement should not be lost as one who really had his eyes on the horizon.

PAGE 79

I heard about this Alan Hodgkin story from a colleague, Vincent Torre, who was a postdoc with Alan Hodgkin. It has since been confirmed in a pub conversation with his son Jonathon Hodgkin, an excellent molecular geneticist at Cambridge. I unfortunately got into science just a tad too late to actually meet Alan Hodgkin, who was ill for the last several years of his life. I am proud to say that I have made the pilgrimage to his laboratory in the basement of the Department of Physiology at Cambridge University. What is most remarkable about it is how unremarkable it is. Great science doesn't always require fancy accommodations.

PAGE 125

The brain—that thing you think you think with—is slightly altered from Ambrose Bierce's definition of the brain in his wry turn of the century, *A Devil's Dictionary* (1906, 1911).

PAGE 174

Dr. Marlys H. Witte, known affectionately as the Ignoramamama, has been using ignorance as an integral tool to teach medical students. Her efforts, begun in 1984, have flowered into an innovative high school science outreach program as well as becoming an integral part of the medical curriculum at the University of Arizona Health Sciences Center. I suggest you have a look at her informative and engaging Web site (http://www.ignorance.medicine.arizona.edu/index.html), which will take you to a publication called *Q-cubed* that describes the innovative programs she and her team have initiated—all based on ignorance as the premiere tool for scientific (and other) education.

Suggested Reading

Barrow, John, *Impossibility: The Limits of Science and the Science of Limits.* Oxford, England: Oxford University Press, 1998. Barrow is a mathematician and theoretical physicist by trade and has written several very readable, popular books on these subjects. He does not oversimplify, which is refreshing. This book muses mostly on the limits of cosmology and what we can know about the universe. It is a rewarding primer on the limitations of our knowledge, which is a kind of knowledge itself.

Chaitin, Gregory, *Conversations with a Mathematician: Math, Art, Science and the Limits of Reason.* New York: Springer, 2002. Chaitin is a mathematician and computer scientist who is one of the leading experts on Gödel and Turing and has proposed some his own complex and provocative theories about information and truth. The essays collected here are among his least technical and most readable.

Crasti, John L., and Karlqvist, Anders, eds., *Boundaries and Barriers: On the Limits to Scientific Knowledge.* New York: Perseus Books, 1996. A collection of relatively short but pointed essays based on talks given at a conference held in the Arctic Circle metropolis of Abisko, Sweden, in 1995. Although Casti and many of the other contributors continue to write and speak and think about these issues, I am not aware of any follow-up conferences or meetings. Seems overdue.

Let me take advantage of this moment to recommend two other less well-known works of Casti, both fiction of a sort. *The One True Platonic Heaven* (John Henry Press, 2003) is billed as "a scientific fiction on the limits of knowledge" and has quite a cast of historical characters making fictional statements—but ones they could have made—regarding what can really be known. The other book, *The Cambridge Quintet* (Perseus Books, 1998) actually came first, but I read it second. This one is subtitled *A Work of Scientific Speculation* and imagines a dinner party hosted by C. P. Snow and attended by Alan Turing, J. B. S. Haldane, Erwin Schrödinger, and Ludwig Wittgenstein. Quite a dinner party. Don't you wish you were there? Get the book.

Duncan, R., and Weston-Smith, M., eds., *The Encyclopedia of Ignorance.* New York: Pergamon Press, 1977. A collection of short articles solicited by the editors for leaders in many fields of science at the time. I have never understood why it was never updated or turned into a regular journal. It's a wonderful idea, although the articles are a bit uneven and could have used some editing. It's still fairly available as a used book, but it's now mostly of historical interest.

Fara, Patricia, *Science: a Four Thousand Year History.* New York: Oxford University Press, 2009. Four thousand years in a little over 400 pages, quite an accomplishment—and Fara doesn't limit herself to Western science (of course, it's only 400 years old). Full of remarkable insights and provocative historical perspectives. I'm not a historian of science, but I bet it's controversial.

Gillespie, Charles Coulson, *The Edge of Objectivity: An Essay on the History of Scientific Ideas.* Princeton: Princeton University Press, 1960. This "essay" goes on for some 550 pages, but the author, a well-known historian of science, writes with such clarity and originality that it is a pleasure to read the book just to be reminded what good writing sounds like. Gillespie successfully blends the historical with the philosophical approaches to understanding how science has come to work the way it does. His knowledge is encyclopedic.

Goldstein, Martin, and Goldstein, Inge F., *How We Know: An Exploration of the Scientific Process.* New York: Da Capo Press (division of Perseus Books), 1981. This husband and wife couple, who both have connections to Columbia University (one went there, the other worked there), have written a wonderfully readable primer on how to be a scientist. In my opinion every undergraduate who thinks they want to be a scientist should be required to read it—no, every undergraduate. Written in 1981, there is not one thing in it that is not relevant today. I found this book by accident on a shelf in the used science book section in the basement of New York's famed Strand bookstore ("18 miles of books"), one of those happy accidents that suggests why we will so miss bookstores.

Gribbin, John, *The Scientists: A History of Science Told Through the Lives of Its Greatest Inventors.* New York: Random House, 2002. John Gribbin is one of the most prolific scientist/science writers on the scene. I personally like this book the best, but perhaps that's because I find it the most scholarly and researched. It is a very straightforward approach to the progress of science in the West, beginning largely in the Renaissance traced through the scientists who did the work—or at least are credited for it. This is of course only one way to trace the progress of ideas, but Gribbin's writing is so clear and the organization so precise that it becomes very insightful.

Gribbin, John, *The Origins of the Future: Ten Questions for the Next Ten Years.* New Haven, CT: Yale University Press, 2006. Gribbin is still one of the most readable scientific writers on the planet. This book poses 10 pretty big questions, mostly cosmological, that I doubt will be solved in the next 10 years. But once again, the writing is crystal clear and you learn more about what's known by looking at what isn't. A formula I can certainly agree with.

Harrison, Edward, *Cosmology: The Science of the Universe,* 2nd edition. Cambridge, England: Cambridge University Press, 2005. A big and unfortunately expensive book that is a technical textbook and at the same time a valuable historical approach to a crucial intersection of physics, mathematics, and astronomy—even some biology appears at the end. A wonderful reference book if this is where your interests are and full of fabulous quotes that Harrison has presumably spent years collecting. He also includes a list of suggested readings at the end of each chapter, which are generally more accessible to a lay audience. His chapter on "Darkness at Night," for example, traces the not-so-obvious solution to an age-old paradox about why in an infinite universe the sky is not ablaze with stars at night. It is a masterful treatment of the process of scientific thinking as well as downright entertaining, even though full of equations.

van Hemmen, J. Leo, and **Sejnowski, Terrance J.,** eds., *23 Problems in Systems Neuroscience.* New York: Oxford University Press, 2006. With a nod to Hilbert, the editors engage a group of modern neuroscientists to write essays about the critical questions that remain in the area of what is called systems neuroscience—that is the study of the brain as a system, and not merely as a collection of parts. Some of the essays stand alone; many require a certain amount of assumed neuroscientific knowledge. Mostly interesting for its emphasis on questions.

Holton, Gerald, *Thematic Origins of Scientific Thought.* Cambridge, MA: Harvard University Press, 1988. Holton is a deservedly well-respected historian of science, and this book traces his particular take on the development of scientific thinking. It is mostly about the physical sciences, and it is a scholarly work meant for the serious student.

Horgan, John, *The End of Science.* Boston: Helix Books/Addison Wesley, 1996. I suppose no list of books on possible limits in science is complete without this popular volume about which much has already been said and written. There are interesting interviews with modern scientists and philosophers of science, a surprising number of whom have passed away since its publication, so it's becoming something of a historical record. The thesis is provocative but almost assuredly wrong, as I suspect the author knows.

Lightman, Alan, *The Discoveries: Great Breakthroughs on 20th Century Science.* New York: Vintage Books (Random House), 2005. Lightman is a physicist turned writer—both fiction and nonfiction, perhaps now best known for his charming book *Einstein's Dreams.* Although largely concerned with the great physics experiments of the day (there are a few chapters on biology and chemistry), Lightman takes the famous papers of Planck, Einstein, Rutherford, Bohr, among others, and helps you read through them. The papers are reprinted, often in slightly abridged format, and are preceded by a deconstruction in lay terms by Lightman, putting the article in historical perspective and alerting the reader what to look out for and what are the key findings. It's a brilliant attempt to allow readers access to what we call the primary literature—that is, the original papers written by the scientists—a territory that is normally way out of bounds for the lay reader. Regardless of whether you care about the particular science being presented (cosmology, astrophysics, relativity, DNA, etc.), this book will allow you to see that you can read these papers, and others as well, and that there is great pleasure to be gotten from doing so.

Lightman, Alan, *A Sense of the Mysterious.* New York: Vintage Books (Random House), 2005. Lightman writes beautifully and in this small book concentrates on the emotional life of the scientists, and how important this often forgotten aspect is to the creative work of discovery. The mysterious is the muse.

Maddox, John, *What Remains to be Discovered.* New York: Touchstone/Simon and Shuster, 1998. The late John Maddox was the editor of *Nature* for a remarkable 23 years, from 1966 to 1973. He oversaw tremendous growth in the number of papers published, and in the influence of the journal and the way science was done. This book is his look into the future through the lens of what we don't know yet. It's a bit of a gamble because it requires predicting as well as questioning, but it certainly does provide plenty of information and plenty of speculation by someone who was at the center of discovery for many years.

Poincare, Hénri, *The Value of Science*. Edited and with Introduction by Stephen Jay Gould. The Modern Library Science Series, New York 2001. Hénri Poincare was a prolific scientist, thinker, and writer who had that knack for explaining difficult things to a nonexpert—whether philosophy to a scientist, or vice versa, or both to a lay public. He has been called the Carl Sagan of his day. This volume brings together his three main books: *Science and Hypothesis*, *The Value of Science*, and *Science and Method*. You buy this as a reference book, and then you find yourself lost in pages full of ideas.

Stanford, P. Kyle, *Exceeding Our Grasp*. New York: Oxford University Press, 2006.

> This is one of the scariest history/philosophy of science books I have ever read. Stanford delves deeply into the problem of "unconceived alternatives"—things we should have thought of, were right in front of our noses, but somehow eluded us, sometimes for centuries. Why the nearly 100-year-long blank between Darwin and the identification of the gene as the basis of heredity? In fact, Darwin had enough information to have postulated the gene, but he never conceived the possibility. The scary point, of course, is what are we looking at and not seeing today?

Additional Articles Consulted

Anderson, P. W. (1997). Is measurement itself an emergent property? *Complexity*, *3*(1), 14–16. doi:10.1002/(SICI)1099-0526(199709/10)3:1<14::AID-CPLX5>3.0.CO;2-E

Casti, J. L. (1997). The borderline. *Complexity*, *3*(1): 5–7.

Caves, C., & Schack, R. (1997). Quantifying degrees of unpredictability. *Complexity*, *3*(1), 46–57.

Chaitin, G. (2006). The limits of reason. *Scientific American, 294*, 74–81.

Daston, L. (1992). Objectivity and the escape from perspective. *Social Studies of Science*, *22*(4), 597–618. doi:10.1177/030631292022004002

Gell Mann, M. (1997). Fundamental sources of unpredictability. *Complexity*, *3*(1), 9–13. doi:10.1002/(SICI)1099-0526(199709/10)3:1<9::AID-CPLX4>3.3.CO;2-I

Glass, D. J., & Hall, N. (2008). A brief history of the hypothesis. *Cell*, *134*(3), 378–381. doi:10.1016/j.cell.2008.07.033

Gomory, R. E. (1995). The known, the unknown and the unknowable. *Scientific American*, *272*(6), 120. doi:10.1038/scientificamerican0695-120

Hut, P., Ruelle, D., & Traub, J. (1998). Varieties of limits to scientific knowledge. *Complexity*, *3*(6), 33–38. doi:10.1002/(SICI)1099-0526-(199807/08)3:6<33::AID-CPLX5>3.3.CO;2-C

Kennedy, D., & Norman, C. (2005). What don't we know? *Science*, *309*(5731), 75. doi:10.1126/science.309.5731.75

Krauss, L. (2004). Questions that plague physics. *Scientific American*, *291*, 82–85. doi:10.1038/scientificamerican0804-82

Maddox, S. J. (1999). The unexpected science to come. *Scientific American*, *281*(6), 62–67. doi:10.1038/scientificamerican1299-62

Schwartz, M. A. (2008). The importance of stupidity in scientific research. *Journal of Cell Science*, *121*(Pt 11), 1771. doi:10.1242/jcs.033340

Siegfried, T. (2005). In praise of hard questions. *Science*, *309*(5731), 76–77. doi:10.1126/science.309.5731.76

Index